# Toward Affordable Systems III

## Portfolio Management for Army Engineering and Manufacturing Development Programs

Brian G. Chow, Richard Silberglitt, Caroline Reilly,
Scott Hiromoto, Christina Panis

Prepared for the United States Army
Approved for public release; distribution unlimited

ARROYO CENTER

The research described in this report was sponsored by the United States Army under Contract No. W74V8H-06-C-0001.

**Library of Congress Cataloging-in-Publication Data** is available for this publication.

ISBN 978-0-8330-6039-6





Published 2012 by the RAND Corporation
1776 Main Street, P.O. Box 2138, Santa Monica, CA 90407-2138
1200 South Hayes Street, Arlington, VA 22202-5050
4570 Fifth Avenue, Suite 600, Pittsburgh, PA 15213-2665
RAND URL: http://www.rand.org/
To order RAND documents or to obtain additional information, contact
Distribution Services: Telephone: (310) 451-7002;
Fax: (310) 451-6915; Email: order@rand.org

# Preface

The third in a series, this monograph expands and applies the RAND Corporation's portfolio analysis and management (PortMan) method to address the problem of selecting Army engineering and manufacturing development (EMD) projects to develop affordable systems in the face of cost and budget uncertainties. Like its predecessors,[1] it focuses on methodology development. While we use data and analysis on capability gaps and near-term systems developed by the U.S. Army Training and Doctrine Command (TRADOC)/Army Capabilities Integration Center (ARCIC), we do so only to provide a more realistic demonstration of the methodology and its applications. Since resource limitations kept us from estimating all input parameters accurately enough to inform decisions about actual EMD projects, readers should not draw conclusions about the merits or drawbacks of any specific projects that we used for demonstration purposes.

This monograph should be of interest to research, development, and acquisition managers who are responsible for portfolio management of programs; engineers in research, development, test, and evaluation programs; and those who are interested in the optimal allocation of funds among different programs and/or developmental stages to fill future capability gaps at the lowest total lifecycle cost.

This research was sponsored by Stephen Bagby, Deputy Assistant Secretary of the Army (Cost and Economic Analysis), Office of Assistant Secretary of the Army (Financial Management and Comptroller), and it was conducted within RAND Arroyo Center's Force Development and Technology Program. RAND Arroyo Center, part of the RAND Corporation, is a federally funded research and development center sponsored by the United States Army. For further information, contact the principal investigators, Richard Silberglitt (email Richard_Silberglitt@rand.org, phone 703-413-1100

[1] Brian G. Chow, Richard Silberglitt, and Scott Hiromoto, *Toward Affordable Systems: Portfolio Analysis and Management for Army Science and Technology Programs*, Santa Monica, Calif.: RAND Corporation, MG-761-A, 2009; Brian G. Chow, Richard Silberglitt, Scott Hiromoto, Caroline Reilly, and Christina Panis, *Toward Affordable Systems II: Portfolio Management for Army Science and Technology Programs Under Uncertainties*, Santa Monica, Calif.: RAND Corporation, MG-979-A, 2011.

extension 5441) or Brian Chow (email Brian_Chow@rand.org, phone 310-393-0411 extension 6719).

The Project Unique Identification Code (PUIC) for the project that produced this document is ASA116025.

For more information on RAND Arroyo Center, contact the Director of Operations (telephone 310-393-0411, extension 6419; FAX 310-451-6952; email Marcy_Agmon@rand.org) or visit Arroyo's website at http://www.rand.org/ard/

## Contents

# Figures

# Tables

# Summary

The U.S. Budget Control Act passed on August 2, 2011, marked a new era of austerity in the nation's budgetary environment. The changes appear to be indefinite and present the Army and the rest of the U.S. Department of Defense (DoD) with unprecedented fiscal challenges. No DoD domain will likely remain untouched, including acquisitions. Yet the U.S. Army's need for mission-capable weapon systems will remain constant. As a result, the Army will need to find ways to ensure that its scientists and engineers are designing both effective *and* affordable systems in this frugal environment.

The 2011 legislation brought added urgency to what had already been a growing premium within DoD: reaping savings through improved efficiency and cost-effectiveness. In 2009, the establishment by Congress of a DoD cost czar to conduct independent cost assessments of new major weapon systems was a portent of this trend. Now, meeting these DoD-wide objectives has become critical. In 2006, Deputy Secretary of Defense Gordon England issued a memorandum requesting that DoD agencies experiment with capability portfolio management for planning and implementing capability development. In a 2008 directive, he formalized his call for experimentation via a mandate that all DoD agencies use capability portfolio management to optimize capability investments and minimize risk in meeting the DoD needs across the defense enterprise.

With these needs and guidance in mind, since 2006, the RAND Corporation has been developing a methodology for selecting and managing portfolios of Army research and development (R&D) projects. Sponsored by the U.S. Army Deputy Assistant Secretary for Cost and Economics, this multiyear work was designed to produce a process for the Army to identify optimal investments in cost-effective, affordable weapon systems. RAND's portfolio analysis and management (PortMan) method and model are the results. RAND has focused each phase of this work not only on developing and refining the methodology, but also on demonstrating PortMan using various portfolios of projects at different developmental stages, so that Army decisionmakers can see it in action at different stages and gain a tangible sense of its value.

## RAND's Latest Work on PortMan Focuses on R&D Projects in the Engineering and Manufacturing Development (EMD) Stage

The defense acquisitions process comprises many stages (see Figure S.1). The earliest of these run in parallel with science and technology (S&T) research. RAND's initial work on PortMan focused on optimizing portfolios in those earliest S&T stages: Companion monographs published in 2009 and 2011 offer demonstrations of PortMan on the Army's highest priority S&T projects, Army Technology Objectives. Yet PortMan can be used equally for portfolios of projects further along the acquisitions pipeline. Most recently, the RAND team turned to analyzing portfolios of projects in, and near, the EMD stage. As Figure S.1 shows, there are narrow windows just before and after EMD populated by near–engineering and manufacturing development (NEMD) projects (i.e., those that are almost ready to enter EMD) and ready-to-be-fielded (RTBF) systems (i.e., finished with EMD, but not yet in full production and deployment). For reasons related to performance and cost-effectiveness, when a planner is prepared to select investments, both NEMD projects and RTBF systems should be considered because they may be profitably substituted for EMD projects. Accordingly, including them in an analysis alongside EMD projects can help optimize EMD portfolios. We refer here

**Figure S.1**
**Stages of the Defense Acquisition Process, with Related S&T Stages**

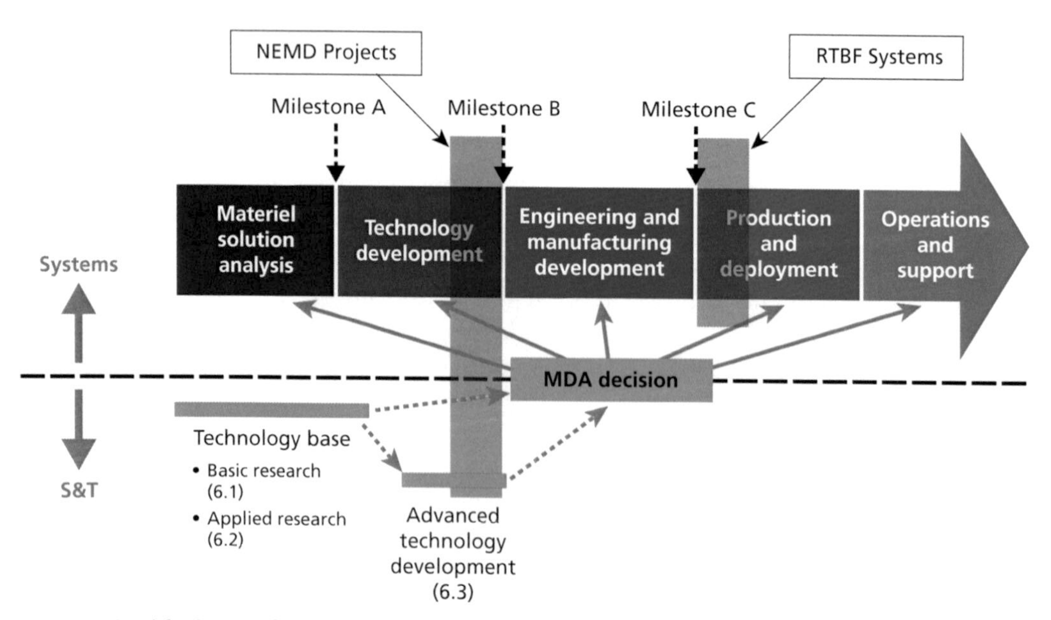


SOURCE: Simplified graph from Bradford Brown, *Operation of the Defense Acquisition System, Statutory and Regulatory Changes*, Kettering, Ohio: Defense Acquisition University, December 8, 2008.
NOTE: MDA = Milestone Decision Authority.
RAND *MG1187-S.1*

to portfolios that include all three groups (N/EMD) as portfolios of EMD projects plus those near EMD's front end (NEMD) and EMD's back end (RTBF).

Together, RAND's two previous PortMan studies and the one we are documenting here provide a comprehensive method and model to analyze portfolios covering all R&D projects and systems up to full production and deployment.

## PortMan Facilitates a New Mindset in Acquisitions Management

The particular challenges the Army faces in selecting projects that will meet future capability needs at an affordable overall cost are not new. Uncertainty has always compounded decisionmaking—uncertainty, for example, about whether all funded projects will succeed, what budgets will be available, and whether changes in the economic or strategic environment will force the Army to alter capability requirements. It is also difficult to think in terms of the "big picture" regarding the full lifecycle costs of fielding a system—from basic research all the way through operations and support—rather than the unique costs of one particular acquisitions stage.

Yet in portfolio analysis, it is important not only to consider the range of uncertainties and full lifecycle costs but also to bring these factors consistently into the evaluation process. Failing to do so can lead to dramatically different, often inferior, investment selections and outcomes—for example, if the result is a false impression of how well a given portfolio fills capability gaps.[2] Today's severe budget constraints also make it advisable to be able to distinguish between threshold ("must have") and objective "desirable" requirements in deciding the extent to fill each gap.

The newest version of PortMan provides a means for the Army's portfolio managers to perform these analyses and more. In the current demonstration with N/EMD portfolios, RAND has designed the process to allow for the possibility that there will be overruns in implementation costs and also that the implementation budget may be less than requested or desirable. With this feature, a manager can identify the best possible portfolio under such uncertainties. RAND has also introduced the concept of *threshold* (must-have) and *objective* (desirable) requirements, rather than a single fixed set of requirements. This feature permits the portfolio manager to measure the costs of aiming to fill all gaps against filling just the essential ones. Managers can also see the variety of anticipated costs when must-have requirements are set at different levels. This allows managers to see the trade-offs between level of requirements met and budget required.

At the same time, this latest version of PortMan retains the key features of previous versions, such as the ability to break out components of the total lifecycle budget

[2] Portfolios of S&T projects meet all capability gaps when the projects have a 100 percent success rate. But they have a very low probability of doing so when uncertainty about whether projects would succeed is taken into account.

and show what percentage of capability gaps can be filled with these component budgets set at various levels. In this way, Army leadership can optimally apportion the total lifecycle budget between costs for R&D and implementation, for example, rather than allocating R&D funds separately on a suboptimal use-it-or-lose-it basis.

Being able to think in terms of overall costs, setting priorities between must-have and desirable capabilities, and bringing uncertainty into the mix can all contribute to a new mindset that would help the Army fulfill the DoD's desire for savings from improved efficiency and effectiveness in a future that realistically will be rife with unknowns.

## Demonstrating PortMan on N/EMD Portfolios Highlights How It Can Assist Army Acquisition Managers

Using a linear programming model together with a simulation, PortMan aids in carrying out two fundamental acquisitions planning tasks:

- **Setting optimal budgets:** In our demonstration, we ask "What is the optimal remaining R&D budget, and the optimal total remaining lifecycle (TRLC) budget for ongoing N/EMD projects?"
- **Selecting an optimal N/EMD portfolio for any set of budgets:** Here we ask "Which N/EMD projects should be terminated, and which continued for any given R&D and lifecycle budgets—including a budget cut?"

To generate the answers to these questions, PortMan also must be able to calculate how well a given portfolio will be able to fill gaps in the Army's current capabilities. One can think of this in terms of a third task: mapping supply (i.e., a particular portfolio) to demand (i.e., given requirements).

### Mapping Supply and Demand

PortMan can evaluate the overall performance of a portfolio of N/EMD projects (the supply) and broadly expose potential problem areas—that is, which requirements (or areas of demand) are at risk of not being met by that particular portfolio. Similarly, PortMan can identify where certain requirements will be met by too many projects, leading to unnecessary redundancies and certain requirements actually being overmet. One can also refer to the results of such an evaluation as the portfolio's "expected value." Because the success of N/EMD projects—that is, whether ultimately they will lead to a fielded system—is uncertain, one must deal with probabilities in assessing their performance. The PortMan concept of *feasible percentage* is a way of managing this inherent uncertainty. *Feasible percentage* indicates the degree of likelihood that a particular portfolio of N/EMD projects within a given budget will meet a set of predefined requirements, including a calculation for the possibility that some projects in the portfolio will ultimately lead to systems with a much higher unit cost than originally expected.

### Setting Optimal Budgets

In our current demonstration, PortMan applies the ability to calculate how well the supply matches the demand to estimating the expected values of various portfolios when budgets are set at different levels. In this way, PortMan can enable portfolio managers to see which budgets will generate a cost-effective portfolio that meets an acceptable level of requirements. It can also help them identify cut-off points below which it would become imprudent to let a budget fall because the chance of meeting requirements would drop to an uncomfortably low level.

**Known-Budget Case.** Figure S.2 provides an illustration of expected values. Here the N/EMD demonstration draws on PortMan's ability to break out components of the total lifecycle budget. For the purposes of the demonstration, we excluded the costs that had already been spent on projects at the time of the analysis—i.e., the sunk costs—and looked specifically at the costs still to be incurred: the total *remaining* lifecycle budget. This total cost includes (1) the remaining R&D cost (the amount required to finish developing a system) and (2) the implementation cost (the cost of acquiring, fielding, operating, and maintaining a system over its lifetime).[3] The Port-

**Figure S.2**
**Likelihood of Meeting Threshold Requirements Within a TRLC Budget for N/EMD Systems**

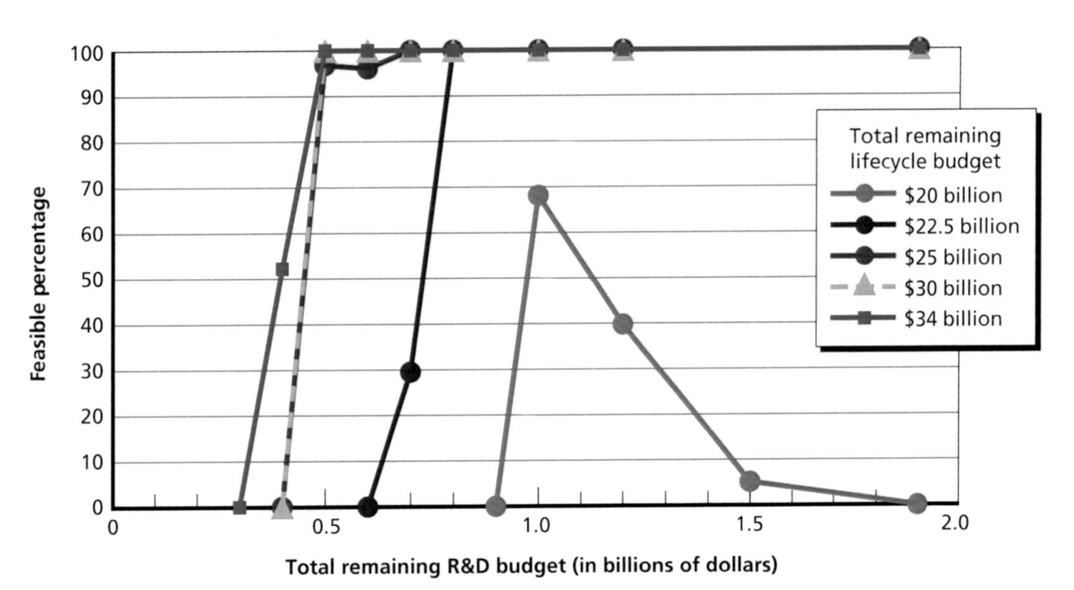


RAND *MG1187-S.2*

---

[3] Here this lifetime is assumed to be 20 years.

Man analysis discloses all three budgets, so that managers can see the trade-offs in setting each of them at different levels.

The demonstration takes five different TRLC budgets and shows (on the y-axis) what percentage of a set of threshold requirements each of those budgets can meet.[4] At the same time, it shows the effects of varying the remaining R&D budget *within* a given TRLC budget. This sort of output can be particularly helpful to portfolio managers in that it reveals a "sweet spot" at which the TRLC and R&D budgets are both at the most cost-effective levels. A closer look at Figure S.2 shows how this plays out—and reveals the degree of influence of the R&D budget amount. As long as the R&D budget remains above a certain amount (here $0.8 billion), there is no difference between setting the TRLC budget at $34 billion, $30 billion, $25 billion, or $22.5 billion: All four alternatives make it fully feasible for the Army to meet all threshold requirements.[5] But even subtle variations in the R&D budget can make a huge difference. For example, when the total lifecycle budget is $22.5 billion, lowering the R&D budget by just $0.1 billion—from $0.8 to $0.7 billion—has dramatic effects, with the probability of meeting requirements plummeting from almost 100 percent to just 30 percent.

In contrast, a total lifecycle budget of $25 billion is more robust to reductions in the total remaining research and development (TRRD) budget. At $0.7 billion for R&D rather than $0.8 billion, the probability of satisfying must-have requirements remains very close to 100 percent. In fact, the R&D budget can fall as low as $0.5 billion with minimal consequences, as the probability of meeting requirements stays above 95 percent. Accordingly, the sweet spot for ongoing N/EMD projects is a TRLC budget of $25 billion and a TRRD budget of $0.7 billion. The sweet spot is the most cost-effective budget with which most of the ongoing projects will be funded. It also suggests that certain projects should be terminated, because the money saved from not funding them can be more cost-effectively spent on new projects.

Because the sweet spot should not be set too close to a point at which the probability of meeting requirements drops off starkly, going lower than $0.7 for R&D is inadvisable. This provides a hedge for unexpected budget contingencies: If, for some reason, it turns out that the R&D budget needs to drop within a window of up to $0.2 billion, the Army's ability to meet requirements will still remain quite strong. Once

[4] A graph that showed the results for objective requirements—that is, the complete set of gaps the Army would ideally like to fill—would likely look very different. For our demonstration, we used capability gaps defined by the Army Training and Doctrine Command's/Capabilities Integration Center as the basis for threshold and objective requirements.

[5] This is most likely because there is enough redundancy between certain projects in these portfolios to overfill capability gaps at all of these budget levels. In addition, recall that the remaining lifecycle budget consists of both the remaining R&D budget and the implementation budget. We assume here in the demonstration that money saved from the total remaining R&D budget will be reapportioned to the total implementation budget, rather than approached from the wasteful use-it-or-lose-it perspective.

PortMan has identified the optimal remaining R&D budget of $0.7 billion, the Army will know that it should plan on needing a total of $24.3 billion to cover the implementation costs of this N/EMD portfolio.[6] In other words, a sweet spot is the budget with which most of the ongoing projects will be funded, suggesting that the money saved from not funding projects that are recommended to be terminated can be more cost-effectively spent on new projects.

**Uncertain-Budget Case.** In a second illustration, Figure S.3 shows the results for a situation in which portfolio managers do not know how much a given budget will be. Here we designate the TRLC budget as the uncertain one. We create a case in the demonstration where this overall budget is equally likely to be $20, $22.5, or $25 billion, but planners cannot know which it will actually turn out to be. We assume that the source of this uncertainty in the overall lifecycle budget is due to uncertainty in how much will be available for the total implementation budget. The results that PortMan generates show portfolio managers why they should not underestimate what will be needed to cover implementation costs, as this is a decisive factor in meeting requirements successfully. Managers should, without question, consider implementation costs when they select which N/EMD projects to keep and which to terminate.

**Figure S.3**
**Likelihood of Meeting Threshold Requirements with an Uncertain TRLC Budget ($20 Billion to $22.5 Billion to $25 Billion Range)**

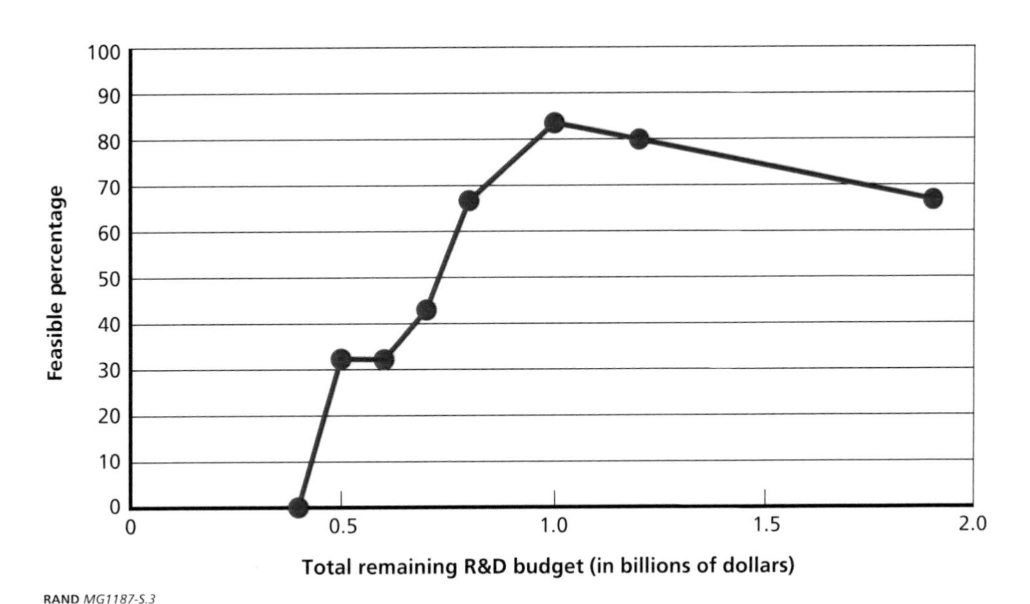


[6] This is because the balance of the total remaining *lifecycle* budget of $25 billion—after apportioning the needed total remaining *R&D* budget of $0.7 billion—is the total *implementation* budget of $24.3 billion.

Under these uncertain conditions, PortMan takes a number of different TRRD budgets and for each and calculates the maximum possible probability that that budget will be able to produce an N/EMD portfolio that will satisfy threshold requirements. Within this demonstration group, PortMan indicates that the best option is the TRRD budget of $1 billion: It gives the Army an 83 percent chance of satisfying must-have requirements. Figure S.3 makes it easy to see that $1 billion constitutes a low-end cut-off point for R&D when the TRLC budget is in this uncertainty range. With any smaller amount, the probability of meeting requirements falls drastically—down by 20 percent even for a relatively small cut of about $0.2 billion. Unlike in the certain budget case, it is not possible to build in a cushion here—staying away from the cliff—because PortMan indicates that a larger R&D budget adds no value in terms of meeting requirements.[7] In fact, Figure S.3 shows that the maximum possible probability of meeting requirements is actually *lower* when the R&D budget exceeds $1 billion.[8]

Our demonstration shows just how much of a difference the presence of uncertainty can make in portfolio planning. Here an uncertain TRLC budget leads to a considerable discrepancy in projected outcomes: The optimal $1 billion R&D budget under heightened uncertainty is *43 percent larger* than the sweet-spot R&D budget of $0.7 billion that PortMan identifies as optimal when the TRLC budget is certain. In brief, PortMan suggests that should portfolio managers face the level of uncertainty specified in our demonstration, they should allocate 43 percent more money for R&D costs. Coping with some uncertainty is nearly unavoidable in planning. Budget cuts, for example—expected or unexpected, for any part of the acquisitions process—are always a possibility. But PortMan can help portfolio managers to handle such uncertainty effectively.

## Selecting an Optimal Portfolio

Within these optimal budgets, PortMan can then recommend which projects to keep in a portfolio and which to drop in order to keep the chances of meeting requirements as high as possible. Countless different portfolios of N/EMD projects are, of course, possible for each combination of remaining lifecycle, R&D, and implementation budgets; various portfolios may perform better or worse. PortMan again draws on its ability to match supply with demand, but now builds in budget as an additional factor: It uses an algorithm that automatically searches for the best combinations of N/EMD projects meeting any given set of budget constraints. The projects that

---

[7] For a given total lifecycle budget, a larger R&D budget will result in less money for implementation.

[8] In Figure S.3, we have assumed that the resources saved by not increasing the R&D budget are diverted instead to the implementation budget.

PortMan selects for continuation are those that will produce systems that will provide the highest chance of meeting all capability requirements under a given budget.

## Recommended N/EMD Portfolio with a Known Budget

To demonstrate, we take the sweet spot of certain budgets from our analysis: $25 billion for lifecycle costs and $0.7 billion for R&D costs. We also again use the threshold requirements. Within these budget constraints, the model points to 17 of the existing 26 N/EMD projects to keep (i.e., those in green) and recommends that the other 9 projects be terminated (i.e., those in red) (see Figure S.4).

The PortMan selection contains some choices that appear on first glance counterintuitive. For example, PortMan flags N/EMD project 84 for continuation, even though its ratio of benefits to R&D cost is similar to many other projects that are rejected. This means that project 84 is contributing to the ability to meet requirements in some important way that other projects are not—and its higher R&D costs are consequently an acceptable trade-off. In contrast, N/EMD project 108 has a benefit-to-cost ratio well within the range of many other projects that are selected. But PortMan nevertheless recommends that it be discontinued. This could be because the system(s) to which it will lead will be redundant with systems generated by other projects, causing certain requirements to be overmet. Or it simply may not contribute to the Army's

**Figure S.4**
**PortMan's Selection of N/EMD Projects with a Known Budget**

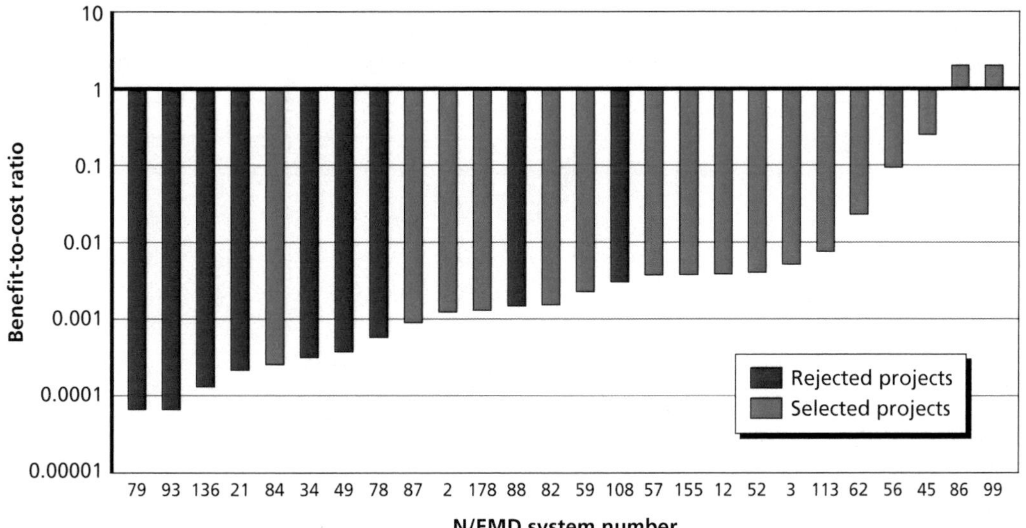


RAND *MG1187-S.4*

ability to satisfy must-have requirements in a significant enough way to warrant continuing it. PortMan flags projects with less value-added in a readily evident way, so that planners can choose the most effective portfolio for the cost.

Our demonstration shows that this kind of analysis—and the portfolio recommendations it produces—is not something that can be replicated by using simple approaches to project selection, such as rank-ordering N/EMD projects according to a basic ratio of total expected benefits to remaining R&D cost. Using the sort of traditional benefit-to-cost-ratio approach rather than the PortMan model would lead to a very different selection. It would first pick the N/EMD project with the highest benefit-to-R&D cost ratio (i.e., the project on the far right in Figure S.4). It would then flag the project with the next highest ratio, and so on, until the TRRD budget was fully committed. But this selection would be suboptimal, resulting in a lower likelihood that the Army would meet all threshold requirements than that possible with the projects in PortMan's recommended portfolio. The extent of sub-optimality depends on the combination of TRRD budget and the TRLC budget. Chapter Four shows three cases where the traditional approach's "optimal" portfolios yield either a 28 percent likelihood to meet all threshold requirements or no likelihood at all, while PortMan's portfolios yield 100 percent, 68 percent, and 40 percent, respectively. These are the results because PortMan's linear programming model and simulation are able to weigh simultaneously the complex interplay of requirements, system capabilities, costs, and uncertainty in a way that no simple criterion or set of criteria can reproduce.

The common use of benefit/cost ratio is not the only problem. We also study the problem of ignoring uncertainties, which PortMan considers. A "certainty model" can use linear programming to address the aforementioned benefit/cost problem. However, for each of its inputs, it would typically use only a single, expected number to approximate a future that is full of uncertainties. Chapter Four shows two of the three cases where a certainty model would produce inferior "optimal" portfolios that mildly lower the likelihood to meet all threshold requirements to 56 percent (from PortMan's 68 percent) and severely to 0 percent (from PortMan's 40 percent). Worse yet, it is not possible to use the certainty model only in the mildly sub-optimal cases because, without PortMan, one cannot tell in advance which cases would yield only mild sub-optimality.

In sum, PortMan finds a portfolio that has a better chance to meet all threshold requirements than those found by models of benefit-cost type and certainty type for any given budget.

## PortMan Offers Army Acquisition Managers a Useful Tool to Balance Effectiveness with Affordability

Today, ensuring that R&D portfolios achieve an optimal balance of effectiveness and affordability in all stages of the defense acquisitions process is essential. RAND's Port-

Man method and model provide novel capabilities that can help the Army fulfill this mission. PortMan offers means to consistently take into account inevitable uncertainties about budgets, the costs of systems, and the performance of projects. Acquisitions managers gain more-realistic assessments of a portfolio's ability to meet requirements and are better equipped to deal wisely with suboptimal budgets or sudden budget cuts. In particular, they can evaluate a very wide range of plausible future budgets and see "cut-off" amounts below which the likelihood of meeting requirements falls rapidly. In addition, PortMan's ability to distinguish between must-have and desirable requirements enables planners to more easily envision a bottom line and make tough but informed choices should budget constraints tighten. By being able to flag projects in a portfolio whose cost outweighs their value, PortMan can potentially help secure sizable long-term cost savings. And by suggesting how to optimally allocate a total lifecycle budget between R&D and implementation, PortMan can help create the new, more cost-effective mindset—toward the use of R&D funds in particular—called for by the current austere budget environment.

# Acknowledgments

We acknowledge the advice, guidance, and financial support of David Henningsen of the U.S. Army Deputy Assistant Secretary of the Army for Cost and Economics, without which this study could not have been possible. We are extremely grateful to Timothy Drake of U.S. Army TRADOC/ARCIC, who provided us with access to the 2007 Capability Gap Analysis and quad charts describing the systems contributing to filling the near-term gaps therein. These provided essential data for the method and model demonstration described in this monograph, which could not have been performed without them. We thank Walter Perry, Steven Popper, and Lance Sherry, whose insightful reviews of the draft manuscript led to substantial improvements in the content and clarity of this monograph. Finally, we thank Bruce Held, our program director at the time this research was conducted, for providing prompt and valuable comments during every stage of the project, from proposal to interim review to final document.

## Abbreviations

| | |
|---|---|
| ARCIC | Army Capabilities Integration Center |
| ATO | Army Technology Objective |
| DoD | Department of Defense |
| EMD | engineering and manufacturing development |
| EV | expected value |
| FOC | force operating capability |
| FP | force protection |
| FY | fiscal year |
| MDA | Milestone Decision Authority |
| N/EMD | near–engineering and manufacturing development and engineering and manufacturing development |
| NEMD | near–engineering and manufacturing development |
| PortMan | (RAND's) portfolio analysis and management method |
| R&D | research and development |
| RDT&E | research, development, test, and evaluation |
| RRD | remaining research and development |
| RTBF | ready to be fielded |
| S&T | science and technology |
| TEV | total expected value |
| TRADOC | Training and Doctrine Command |
| TRLC | total remaining lifecycle |
| TRRD | total remaining research and development |

CHAPTER ONE

# Introduction

This monograph describes the results of the third in a series of RAND studies aimed at developing methodology for selecting and managing portfolios of research and development (R&D) projects. This series of studies, sponsored by the U.S. Army Deputy Assistant Secretary for Cost and Economics, focuses on providing the Army with a method and approach to identify investments in cost-effective and affordable mission-capable weapon systems. This focus is in keeping with recent trends in the Department of Defense (DoD) that emphasize the importance of cost-effectiveness and portfolio management.[1]

As in the two companion studies,[2] this methodology provides a way to manage the relationships between performance requirements, system/technology capabilities, and costs to maximize the reliability of delivering the required overall/systemwide capabilities within a given budget. The idea of reliability is accounted for by the treatment of uncertainty. Building on the methods developed during the previous studies, we describe in this monograph new quantitative tools for capability portfolio management that allow trade-offs between capabilities that fill defined gaps and the R&D cost and lifecycle cost to acquire those capabilities. In this analysis, we exclude all past expenditures (sunk costs) to provide a means for assessing the investments that must be made now and in the future to achieve specific capabilities. As shown in Table 1.1, (remain-

---

1 For example, the Weapon Systems Acquisition Reform Act of 2009 (see P.L. 111-23, Weapon Systems Acquisition Reform Act of 2009, "Title II, Acquisition Policy," May 22, 2009) mandated independent cost assessments of major new military systems and evaluation of trade-offs between cost, schedule, and performance to ensure affordability, and DoD Directive #7045.20 (DoD, Deputy Secretary of Defense Gordon England, *Capability Portfolio Management*, DoD Directive #7045.20, September 25, 2008) mandated the use of capability portfolio management to optimize defense capability investments. See also DoD, Deputy Secretary of Defense Gordon England, *Capability Portfolio Management Test Case Roles, Responsibilities, Authorities, and Approaches*, September 14, 2006. The Vice Chief of Staff of the Army has implemented the latter via an annual review process for 17 different capability portfolios.

2 Brian G. Chow, Richard Silberglitt, and Scott Hiromoto, *Toward Affordable Systems: Portfolio Analysis and Management for Army Science and Technology Programs*, Santa Monica, Calif.: RAND Corporation, MG-761-A, 2009 (hereafter, *TAS-1*); Brian G. Chow, Richard Silberglitt, Scott Hiromoto, Caroline Reilly, and Christina Panis, *Toward Affordable Systems II: Portfolio Management for Army Science and Technology Programs Under Uncertainties*, Santa Monica, Calif.: RAND Corporation, MG-979-A, 2011 (hereafter, *TAS-2*).

**Table 1.1**
**Definitions of Cost Components Used in This Monograph**

| Cost Component | Definition |
|---|---|
| Remaining R&D cost | Remaining cost to complete all stages of R&D, consisting of the S&T stages: 6.1 Basic Research, 6.2 Applied Research, and 6.3 Advanced Technology Development; and the Acquisition System's stages up to and including the Army's Engineering and Manufacturing Development stage |
| Implementation cost | The cost to acquire, field, operate, and maintain systems that are derived from the completed R&D over its life (assumed for the purposes of this monograph to be a 20-year planning period) |
| Remaining lifecycle cost | The sum of the remaining R&D cost and implementation cost |

ing) lifecycle cost of a system consists of two components: (1) (remaining) R&D cost and (2) implementation cost. Remaining R&D cost is defined as the remaining cost to complete the development of the system. The implementation cost is then the cost to acquire, field, operate, and maintain the system over its life (assumed for the purposes of this monograph to be a 20-year planning period). We use a linear programming model together with a simulation[3] to select project and system portfolios that maximize the probability of meeting defined capability gaps for specific values of total portfolio R&D cost and total portfolio implementation cost.[4] This approach allows one to select a portfolio of projects that is most robust (i.e., likely) to fulfill capability gaps in the face of future uncertainty.

## This Study Covers Three Types of Developing Systems

DoD and Army capability portfolios contain projects and systems at different stages of R&D, or stages of the Defense Acquisition System, respectively. Figure 1.1, from a recent Defense Acquisition University briefing, schematically illustrates the relationship between the science and technology (S&T) stages and various stages of acquisition. As shown in the bottom half of the figure, S&T encompasses the 6.1 and 6.2 stages of R&D (often referred to as the technology base), as well as the 6.3 stage, Advanced Technology Development. The top half of the figure shows the stages of the Defense Acquisition System, including engineering and manufacturing development (EMD). While the natural progression of an S&T project would be into the Acquisition System's technology development or the EMD, the figure indicates how S&T projects can transition into any stage of acquisition, based on the decision of a Milestone Decision Authority (MDA). Such an MDA decision would be based on whether

[3] This study uses Monte Carlo simulation.

[4] The probabilistic approach is necessary to reflect the reality that implementation cost and future budgets are uncertain.

**Figure 1.1**
**Relationship Between the Defense Acquisition System and the S&T Stages**

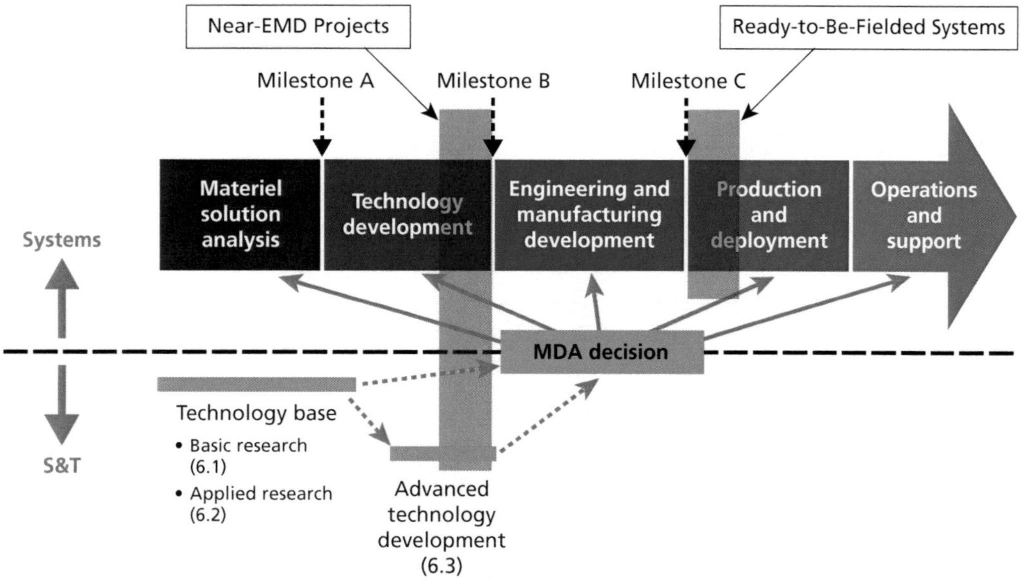

SOURCE: Simplified graph from Bradford Brown, *Operation of the Defense Acquisition System, Statutory and Regulatory Changes*, Kettering, Ohio: Defense Acquisition University, December 8, 2008.
RAND *MG1187-1.1*

the project to be transitioned is capable of passing one or more of the acquisition milestones, each of which has its own specific requirements.

Our previous studies, described in TAS-1 and TAS-2 cited above, developed methods for selecting portfolios of S&T projects[5] that lead specific systems to meet capability gap requirements at minimum lifecycle cost of the systems developed from the completed R&D and within S&T budget constraints.[6] According to Figure 1.1, this is equivalent to selecting portfolios of R&D projects up to and including the technology development stage of acquisition. The study described in this monograph complemented these previous studies by developing methods for portfolio analysis of projects in the EMD stage.[7] The three studies together thus provide methods to ana-

---

[5] The studies dealt exclusively with Army Technology Objectives (ATOs), the Army's highest priority S&T projects.

[6] In these earlier studies, the R&D budget constraint was on total remaining S&T cost because we were interested in selecting the best portfolio of S&T projects for continued funding. EMD cost was included in the implementation cost.

[7] For which EMD cost is part of remaining R&D cost because, in this study, we are interested in selecting the best portfolio of EMD projects, as well as projects near the EMD stage for funding.

lyze capability portfolios encompassing the full range of projects up to, but not including, production and deployment.

The vertical bars in Figure 1.1 highlight two narrow portions of the acquisition system that are of special importance to this monograph: (1) R&D projects that are almost ready to enter the EMD stage, which we define as near–engineering and manufacturing development (NEMD) projects, and (2) systems that have completed their EMD and are ready to be fielded (RTBF) but have not yet entered into full production and deployment. Both the NEMD projects and the RTBF systems represent alternative investments to those in EMD projects. NEMD projects may be preferred over EMD projects because, for certain capability gaps, the Army may be willing to wait the additional time that the R&D will take because the system derived from the R&D may be superior in performance or cost and existing capabilities can fulfill the mission during the extra development time. In other cases, RTBF systems may provide a more cost-effective way to deliver capabilities because the R&D costs of these systems, as opposed to those to be derivable from EMD, have already been fully paid. Thus, analysis of EMD portfolios must be made under the consideration that these two types of alternative investments are potentially substitutes for EMD investments.

## Comparison of the Three RAND Studies

The study described in this monograph developed portfolio management methods and tools for answering two basic questions: (1) what total remaining research and development (TRRD) budget is optimal for the currently ongoing EMD projects? and (2) which existing EMD projects should be terminated and which should be continued for any given budget, including budget cuts? As shown in Table 1.2, there are similarities and differences between this and the previous studies. All three provide a process for portfolio management that includes models and simulations. Moreover, all three focus on new systems, as well as new or improved subsystems placed on legacy systems.[8] All measure the cost of the new systems or improved subsystems marginally, relative to the cost of the legacy systems they are replacing.[9] All measure the value of the new systems in terms of their contributions to meeting capability gaps that exist in the presence of the legacy systems.

The study described in this monograph differs from the previous studies both in its focus on NEMD projects, EMD projects, and RTBF systems and by adding the following new features to our portfolio analysis and management model: (1) allowing

---

[8] We define legacy systems as currently fielded weapon systems and additional units of the current systems to be fielded in the future.

[9] Thus allowing us to capture cost savings through a negative cost for cheaper systems that perform the same (or improved) functions as the legacy systems they replace.

**Table 1.2**
**Factors Incorporated into the Studies**

| | TAS1 | TAS2 | TAS3 |
|---|---|---|---|
| Lifecycle cost at an early stage | ✓ | ✓ | ✓ |
| Not all projects are successful | | ✓ | |
| Possibility of implementation cost overrun | | | ✓ |
| Possibility of implementation budget cuts | | | ✓ |
| Distinction of threshold and objective requirements | | | ✓ |
| Focus on ATOs | ✓ | ✓ | |
| Focus on EMDs, NEMDs, and RTBFs | | | ✓ |

for the possibility of implementation cost overruns; (2) allowing for uncertainty in the budget; and (3) allowing for threshold ("must have") and objective ("desirable") requirements, rather than fixed requirements. The first two new features will allow the portfolio manager to select optimum portfolios based on the realities that many systems incur cost overruns and budgets are often less than requested or desirable, by incorporating these uncertainties in the optimization procedure. The third new feature will allow the portfolio manager to see the trade-off of setting the must-have requirements at various levels and the different costs expected to be incurred in order to meet requirements at these levels. On the other hand, while, like the second study, the version of our model used in this study can accommodate project failure, we ignore this feature to focus the analysis on budget uncertainties and cost overruns.

## Outline of This Monograph

In Chapter Two, we describe the elements of the methodology developed during this study. In Chapter Three, we describe how we estimate the values contributed by each system to individual capability gap requirements. We do this for systems derived from NEMD or EMD projects, as well as for RTBF systems. In Chapter Four, we demonstrate the methodology by applying it to portfolio management of developing systems in the NEMD and EMD stages. This analysis includes full consideration of RTBF systems by explicitly assessing how their presence would affect the selection of NEMD and EMD projects for continued funding. Chapter Five addresses the possibility that performance and cost data for RTBF systems may not be available or properly updated. Accordingly, we develop an approximate consideration methodology to select NEMD and EMD projects for continued funding under these conditions. Chapter Six provides a summary of our findings and recommendations.

CHAPTER TWO

# Description of the Method

This monograph addresses two important portfolio management issues for NEMD and EMD projects,[1] which we will call N/EMD projects hereafter: (1) the optimum size of the budget for ongoing N/EMD projects and (2) which N/EMD projects should be terminated and which continued for any given N/EMD budget. As noted previously, it is important to include consideration of RTBF systems[2] when addressing these issues, because fielding RTBF systems is an alternative investment to developing and fielding systems that will be derived from completed N/EMD projects. However, it is sometimes the case that necessary RTBF system cost and performance data are not available. Accordingly, we describe below two versions of our method. The first includes full consideration of RTBF systems, while the second uses an approximation based on incomplete RTBF system data. This chapter describes our method, while Chapters Three through Five detail an example application that uses data and analysis from the 2007 Training and Doctrine Command (TRADOC)/Army Capabilities Integration Center (ARCIC) capability gap analysis to define requirements and estimate the values of N/EMD-derived and RTBF systems to meet capability gap requirements.

## N/EMD Portfolio Management with Full Consideration of RTBF System Data

Our method is executed in the following series of steps:[3]

1. Determine the requirements to be met by the portfolio.

[1] As defined in the previous chapter, EMD projects are those in the engineering and manufacturing development stage of the Defense Acquisition System, and NEMD projects are those that are near to, but have not yet entered, the EMD stage. See Figure 1.1.

[2] Systems that have completed EMD and are ready to be fielded, but for which a decision for full production and deployment has not yet been made. See Figure 1.1.

[3] The modifications of the method for N/EMD portfolio management with approximate consideration of RTBF system data are described in the beginning of Chapter Five.

2. Define the set of N/EMD projects and RTBF systems to be considered for portfolio selection.
3. Estimate the expected value (EV)[4] contributions of each N/EMD project and RTBF system to meeting the requirements.
4. Estimate the remaining R&D cost of each N/EMD project.
5. Estimate the remaining lifecycle cost of each N/EMD project and RTBF system.
6. Select the N/EMD projects that provide the optimum portfolio for starting or continuing EMD.

### Step 1: Determine Requirements

We adopt as our requirements for this demonstration of our model the 2007 Army capability gaps defined by TRADOC/ARCIC.[5] In its capability gap analysis, TRADOC/ARCIC defined both near-term and longer-term residual gaps. In our previous monographs, TAS-1 and TAS-2, we demonstrated our model using portfolio analysis of Army S&T projects (specifically, ATOs) that address the longer-term residual gaps. Here we analyze N/EMD portfolios aimed at addressing the near-term gaps. Moreover, on June 4, 2010, Defense Secretary Robert Gates stated his intent to save $100 billion from the five-year defense budget (fiscal years [FYs] 2012–2016).[6] In this climate, the Army and the other services will need to take a hard look at required capabilities. Thus, we suggest consideration of two levels of requirements: a minimum threshold that still must be met under an austere budget and an objective requirement that is desirable to meet if adequate resources are available. For our demonstration case below, we will use the TRADOC/ARCIC–defined near-term capability gaps as the objective requirements and assume that the threshold requirements are 75 percent of these objectives.[7]

### Step 2: Define Projects and Systems to Be Considered

The criterion for consideration of N/EMD projects and RTBF systems in our model demonstration is that systems derived from these N/EMD projects after they are completed (hereafter, N/EMD-derived systems), or from these RTBF systems, can be fielded soon enough to meet the near-term capability gap requirements defined by TRADOC/ARCIC.

---

[4] In TAS-1 and TAS-2, we define the value contribution to meeting a requirement as a random variable, with its EV determined by the probability distribution of its possible states.

[5] Tim Drake, TRADOC/ARCIC Asymmetric Warfare Division, private communication. In this monograph, we list gaps and systems by number only. An Excel spreadsheet that identifies the gaps and systems is available, upon approval by the sponsor of this study. If interested, please see the Preface for author contact information.

[6] Office of the Assistant Secretary of Defense (Public Affairs), *Secretary of Defense Provides Guidance for Improved Operational Efficiencies* and *Fact Sheet: Savings and Efficiencies Initiative*, June 4, 2010a and 2010b.

[7] One can also use different sets of threshold requirements and study the needed budgets to meet these sets of requirements. One can then see the trade-off between requirements and budget.

**Step 3: Estimate Expected Values**

Following TAS-1 and TAS-2, we define the contribution of N/EMD-derived or RTBF system $i$ to capability gap $j$ as $V_{ij}$, a random variable whose randomness is described by the probability distribution, $P_s(V_{ij})$. Since the contribution of N/EMD-derived or RTBF system $i$ can end up at a number of states where $s = 1, 2, \ldots S$, the EV of $V_{ij}$ is

$$E[V_{ij}] = \sum_{i=1 \text{ to } s} P_s(V_{ij})V_{ij}.$$

Since some studies may use a Delphi or analytical method to estimate the $E[V_{ij}]$ directly, it can also be called the EV score, or EV, for the $i$th system on the $j$th capability gap. It is this EV that we mean when we use the term *expected value* in this monograph.

In our previous studies described in TAS-1 and TAS-2, we developed a method for estimating the EV of each R&D project that could potentially be selected for the portfolio. This method bases EVs on estimates of how well systems developed from successful projects fill the longer-term residual capability gaps defined by TRADOC/ARCIC. In the current study, with near-term gaps as requirements, we made use of the fact that TRADOC/ARCIC's Capability Gap Analysis VI identifies systems derived from N/EMD projects and RTBF systems that contribute to each capability gap.[8] TRADOC/ARCIC also provides estimates of the level to which these contributing systems combined fulfill each capability gap requirement. In Chapter Three, we show how to use these TRADOC/ARCIC estimates to determine the total EV score for each capability gap and describe a method to disaggregate these scores into EVs for each system that contributes to each capability gap. As with our earlier work, this method can be used by analysts or by a Delphi panel of subject matter experts.[9]

**Step 4: Estimate Remaining R&D Costs**

As shown in Table 1.1, in this study we defined a system's remaining R&D costs to include future development costs through the EMD stage. Analyses performed as input to Army project selection typically consider the candidate systems' remaining R&D costs, and Army portfolio managers are required to submit estimates of these costs for

---

[8] Our logic is to use the N/EMD and RTBF systems, which can be deployed sooner than S&T-derived systems, to meet near-term gaps. By the same logic, we used farther-away S&T-derived systems to meet longer-term gaps, as well as near-term gaps that cannot be met by N/EMD and RTBF systems.

[9] In TAS-1 and TAS-2, we determined the EVs through analysis. For examples of demonstrations of RAND's PortMan method using Delphi panels, see Richard Silberglitt, Lance Sherry, Carolyn Wong, Michael Tseng, Emile Ettedgui, Aaron Watts, and Geoffrey Stothard, *Portfolio Analysis and Management for Naval Research and Development*, Santa Monica, Calif.: RAND Corporation, MG-271-NAVY, 2004; or Eric Landree, Richard Silberglitt, Brian G. Chow, Lance Sherry, and Michael S. Tseng, *A Delicate Balance: Portfolio Analysis and Management for Intelligence Information Dissemination Programs*, Santa Monica, Calif.: RAND Corporation, MG-939-NSA, 2009.

review and approval. Thus, such input data should be readily available to Army portfolio analysts who wish to use our model to select an optimal portfolio of projects.

For this study, TRADOC/ARCIC provided us with quad charts for the N/EMD projects,[10] some of which show remaining R&D costs. For the missing data, we used a publicly available source of R&D cost data: the Research, Development, Test and Evaluation (RDT&E) Budget Item Justification Sheets (R-2 Exhibits) that are required for RDT&E projects requesting more than $10 million in any FY. Each sheet describes actual past and estimated future RDT&E budgets of a program element, as well as its multiple sub-elements, by FY. It is possible to directly match many of our N/EMD projects that have missing R&D cost data with a sheet or a program element based on project and program titles. However, for some projects, overlap between projects and program elements made it impossible to precisely determine which portion of the budget should be attributed to a given project. In the demonstration of our model described in the following chapters, we used our best judgment in such cases.

### Step 5: Estimate Remaining Lifecycle Costs

As shown in Table 1.1, a system's remaining lifecycle cost consists of two components: the remaining R&D cost and the implementation cost. The latter consists of the costs for acquiring, operating, and maintaining the number of systems needed to meet requirements. For major weapon or supporting systems, the program manager is required to estimate and report the remaining lifecycle costs for milestone A approval (see Figure 1.1). Thus, when the project passes milestone B and enters into EMD, estimates of these costs should have already been made. On the other hand, the program manager is not required to submit remaining lifecycle cost estimates for less-expensive systems. In fact, the S&T community has often deemphasized the implementation part of the remaining lifecycle costs, focusing instead on the R&D cost and the potential value of the system. Some S&T planners have also argued that implementation cost is difficult to estimate accurately, even in the later stages of S&T, and that any critical cost issues can be taken into account in value estimates when ranking projects.

On the contrary, we believe that cost estimates should not be mingled with value estimates, since they measure fundamentally different factors.[11] As shown in TAS-1 and TAS-2, it is possible to take account of implementation cost explicitly in portfolio analysis. Our method and model are designed to accept value and cost estimates as independent inputs and to select cost-effective portfolios that consider trade-offs based on both value and cost. As illustrated in TAS-1, TAS-2, and Chapter Four, implementation cost can have an important effect on portfolio selections. In fact, we show in Chapter Four that project rankings based on simple benefit-cost ratios do not provide

---

[10] TRADOC/ARCIC also provided us with quad charts for the RTBF systems.

[11] *Value* is a measure of effectiveness, while *cost* is the expense to obtain that value. In typical cost-effectiveness metrics, value occurs in the numerator and cost in the denominator.

optimal portfolios. In Chapter Four, we also include uncertainty in the unit cost of systems, which can lead to implementation cost overruns. The current DoD and Army focus on affordability underscores the importance of including such uncertainty in Army portfolio analysis.

### Step 6: Select Optimum Portfolio

An optimal portfolio is one that best meets an objective. Our approach uses two models. The linear programming model is a certainty model, and its objective is to find the portfolio that calls for the lowest budget to meet all requirements and constraints. We also use a simulation model to account for uncertainties. Its objective is to find a portfolio that has the highest chance (i.e., feasible percentage) to meet all requirements and constraints.

TAS-1 described the use of a linear programming model to optimize portfolios for affordability, while TAS-2 added a simulation model to take into account the possibility of R&D project failure. In this study, we used both the linear programming model and the simulation model to optimize portfolios under two different assumptions: (1) a known total remaining lifecycle (TRLC) cost budget and (2) an uncertain TRLC cost budget. For case 2, we assumed an equal probability of one-third for three different budget levels—the TRLC budget determined by our model to be most cost-effective for the ongoing N/EMD projects, a lower budget, and an even-lower budget.[12]

For either assumption shown above, the linear programming model and the simulation model can be used together to determine which of the ongoing N/EMD projects should continue to be funded to achieve the highest likelihood of meeting all threshold requirements at a given TRRD budget and TRLC cost budget. One can also determine the optimal split of the TRLC budget into TRRD budget and implementation budget that yields the highest likelihood of meeting all threshold requirements.

## N/EMD Portfolio Management Using Incomplete RTBF Data

Sufficient data, especially implementation cost estimates, for some RTBF systems are not available or updated comprehensively enough to carry out the procedure described in the previous section. We recommend that the Army regularly develop and update performance and cost data for RTBF systems, both to allow their use in the capability portfolio review process and to support budgetary decisions. However, recognizing that this process may take time to develop, we describe below a method for the selection of N/EMD projects for continued funding using incomplete RTBF system data.

[12] For the case in Chapter Four, we used 10 percent and 20 percent for the lower budget and the even lower budget, respectively. For the case in Chapter 5, we used about 15 percent and 30 percent instead. These numbers are not identical because the reference budgets for the two cases are different: $25 billion in Chapter Four and $7 billion in Chapter Five.

The first step in this approximate consideration method is to estimate the contributions of systems derived from N/EMD projects to meeting requirements, such as the capability gaps identified by TRADOC/ARCIC.

The second step is to study whether any of these N/EMD-derived system contributions can be made more cost-effectively by RTBF systems. Clearly, one cannot accurately determine RTBF system cost-effectiveness without accurate RTBF system cost estimates. However, a workable proxy can be obtained in some cases by comparing the engineering design of a developing system at the N/EMD stage with the designs of similar RTBF systems, taking into account the fact that the RTBF systems do not require additional R&D.

The third step is to see whether there are projects that have not yet reached the N/EMD stage but could potentially lead to systems that contribute to capability gaps for which immediate filling is not essential. Subtracting out the contributions from systems identified here and in the previous step provides the objective requirements for the N/EMD systems to meet.

The fourth and final step is to determine the threshold or minimum requirements for the N/EMD systems to meet, which must be based on the latest updated DoD objectives and strategy. In Chapter Five, we use 75 percent of objectives as the threshold requirements for the numerical demonstration of the method.

CHAPTER THREE

# Requirements, Systems, and Expected Value Estimates for Model Demonstration

In its 2007 current force capability gap analysis, TRADOC/ARCIC defined capability gaps in ten different areas. Based on resource availability, and because our purpose was to demonstrate the model, not to select actual projects for continued funding, we chose to limit the scope of this study to the first of these ten areas, force protection (FP).[1] Within the FP area, TRADOC/ARCIC identified 31 capability gaps, 4 NEMD-derived systems, 22 EMD-derived systems, and 157 RTBF systems that contribute to filling these capability gaps. These capability gaps and systems provided the requirements (the demand) and systems to be considered (the supply) for the portfolio optimization demonstration described in this monograph.

## Development of an Expected Value Scale

The RAND PortMan[2] method described in this monograph is derived from a decision framework that estimates EVs using concepts of decision theory together with metrics and scales based on the best available data and expert analysis.[3] In TAS-1 and TAS-2, the requirements were gaps in Army force operating capabilities (FOCs), and we developed metrics and scales and estimated EVs in terms of how well the capabilities provided by systems derived from ATOs matched the capability gaps within each FOC. This "bottom up" method allowed us to estimate the EVs for each ATO and gap and to develop the EV matrix needed for portfolio optimization.

TRADOC/ARCIC's 2007 capability gap analysis, in addition to providing the capability gap requirements and systems to be considered, provided a "Near-Term Rating—a subjective assessment of how near-term solutions mitigate sub-capability

---

1 *Force protection* is a generic term. It is not the title given to this area by TRADOC/ARCIC.

2 PortMan stands for portfolio analysis and management. It is a term we use to describe the method and model we develop and apply in this area.

3 Richard Silberglitt and Lance Sherry, *A Decision Framework for Prioritizing Industrial Materials Research and Development*, Santa Monica, Calif.: RAND Corporation, MR-1558-NREL, 2002.

gaps."[4] Here the near-term solutions are systems we are considering, and the sub-capability gaps are the near-term capability gaps we are using as requirements. TRADOC/ARCIC's near-term rating (for the aggregated systems) reflects the extent to which all the N/EMD-derived and RTBF systems together can meet the requirements. The following definitions are used for the rating:

- Red—does not (even partially) enable mission performance to standard (in the near term).
- Amber—can partially enable mission performance (to standard in the near term).
- Green—enables mission performance to standard (in the near term).

Since requirements with a near-term rating of red cannot be met in the near term, they would have to be met in the longer term. In other words, red-rated requirements should be addressed by S&T projects rather than the near-term N/EMD projects we are considering. Figure 3.1 shows the distribution of TRADOC/ARCIC's near-term ratings for the 31 FP capability gaps. None of the gaps received a green rating, nine received a red rating (and thus will not be considered further in this monograph), fifteen received an amber rating, and seven a rating of amber/red. The assignment of amber/red implies that TRADOC/ARCIC staff assessed these requirements as being partially met, but at a level less than that implied by a rating of amber. Accordingly,

**Figure 3.1**
**Distribution of TRADOC/ARCIC Near-Term Ratings for FP Capability Gaps**

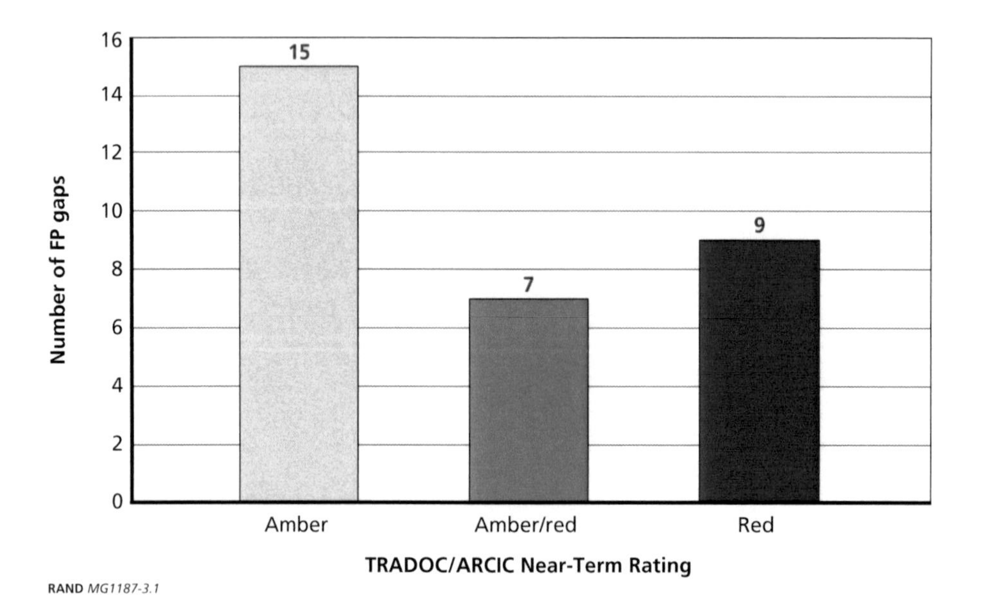


RAND *MG1187-3.1*

[4] Tim Drake, TRADOC/ARCIC Asymmetric Warfare Division, private communication.

when developing a scale for EV below, we assign amber/red a value halfway in between that of amber and that of red.

TRADOC/ARCIC identified 183 systems that provided capabilities to fill FP capability gaps and thus contributed to these near-term ratings. Figure 3.2 shows the distribution of these systems by stage of development. Most of the systems (157 out of 183) are RTBF systems, which raises the question of whether it is a good strategy for the Army to fund so many of its R&D projects all the way to the final stage of system development. A large pool of RTBF systems does provide the Army with a more rapid fielding capability in the face of a rapidly changing threat environment, e.g., as it has been facing in recent years in Iraq and Afghanistan, especially with respect to asymmetric threats such as improvised explosive devices. On the other hand, even a large pool of RTBF systems might still not be prescient enough to meet many of the threats that actually emerge. Even if they were, the developed system might not be able to be massively deployed quickly enough to meet the threat. Too many RTBF systems could consume scarce R&D resources that might otherwise fund N/EMD projects to develop other needed capabilities. Moreover, if many of the RTBF systems are never fielded, those R&D funds might be considered to have been wasted. This important issue about the balance between funding for N/EMD projects and RTBF systems has added currency and importance in the current stressed budget situation. Because this issue deserves a dedicated study on its own, we do not address it in this monograph. Rather, we take the systems and requirements data shown in Figures 3.1 and 3.2 as

**Figure 3.2**
**Distribution of N/EMD-Derived, EMD-Derived, and RTBF Systems Contributing to FP Capability Gaps by Stage of Development**

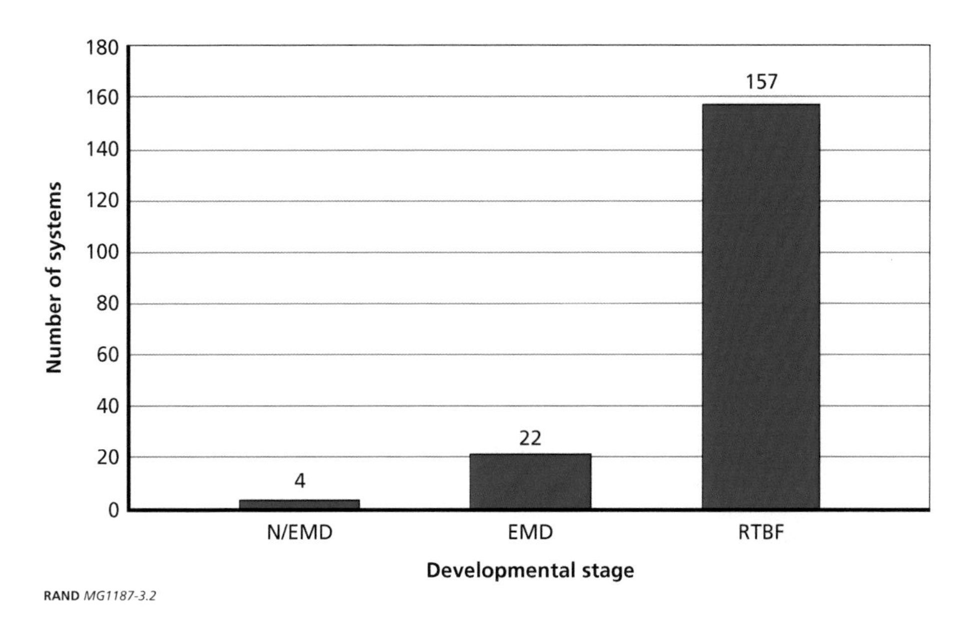


RAND *MG1187-3.2*

given and use them for portfolio optimization to demonstrate our model. In other words, we do not question whether the Army funds were efficiently used to produce this many RTBF systems. Rather, we consider, given the situation shown in Figure 3.2, how the Army should select a subset of N/EMD projects for further development under a given total R&D and total lifecycle budget in order to take full advantage of the RTBF systems already developed.

To develop an EV scale based on the TRADOC/ARCIC near-term ratings, we note that the definitions of *red* and *green* correspond to zero and unity, respectively, if the scale is based on mission performance to standard. The TRADOC/ARCIC definition of amber by itself does not allow its placement on this scale. However, TRADOC/ARCIC used two additional ratings, the amber/red rating shown in Figure 3.1 and amber/green for specific gaps. Thus, this scale suggests the following five-point EV scale as a reasonable representation of the TRADOC/ARCIC near-term ratings:

- Red corresponds to EV = 0.
- Amber/red corresponds to EV = 0.25 or 25 percent.
- Amber corresponds to EV = 0.5 or 50 percent.
- Amber/green corresponds to EV = 0.75 or 75 percent.
- Green corresponds to EV = 1 or 100 percent.

Use of this EV scale determines the total (aggregated) expected value contributions from the 183 systems identified by TRADOC/ARCIC for each of the 22 (non-red) FP capability gaps, as shown in Figure 3.3. However, this "top down" EV determination does not provide the individual system EVs that are needed for portfolio optimization. In the following section, we describe our method for disaggregating these total EVs into the estimated EVs for each individual N/EMD-derived and RTBF system.

## Disaggregation of Total EV Contributions to Estimate EVs of Individual Contributing Systems

To estimate the individual system EVs, we first allocated the systems to the FP capability gaps to which they contribute, as determined by TRADOC/ARCIC. We then executed the following six-step procedure to determine the EV of each project or system for each of the 22 FP capability gaps that were not given a near-term rating of red by TRADOC/ARCIC.

**Figure 3.3**
**Total EV Contributions to FP Capability Gaps Derived from TRADOC/ARCIC Near-Term Ratings**

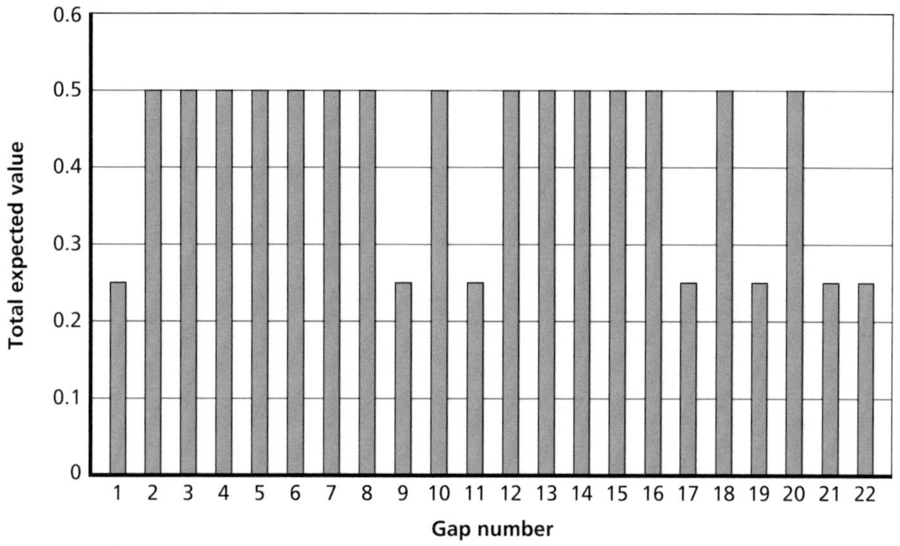

RAND *MG1187-3.3*

### Step 1—Determine Systems to Participate in Disaggregation

In most cases, we found sufficient data on system objectives, schedule, and performance to allow us to estimate their relative contributions to the EVs shown in Figure 3.3.[5] However, for a small number of systems, we judged the data to be insufficient, and these systems were not included in the disaggregation. There were also some cases in which the interdependencies between systems were so strong that we judged it was necessary to treat these as a group when disaggregating their contributions.

### Step 2—Group Systems According to Approaches

To compare the relative importance of systems in contributing to filling FP capability gaps, we put them into groups according to the approaches used to provide the needed capability. This provided an organizational structure and a means to initially compare the various alternative approaches to achieving a capability, which was a much more tractable problem than direct evaluation of the contributions of a large number of individual systems. In the following hypothetical example to illustrate the method, if the capability sought was improved body armor, the groups might be fiber-based soft systems, ceramic-based hard systems, or hybrid systems that include both fibers and ceramics. This grouping approach simplifies the estimation of relative EVs because,

[5] For most systems, these data were contained on quad charts provided by Tim Drake, TRADOC/ARCIC Asymmetric Warfare Division. We supplemented these quad charts where necessary with data publicly available on the Internet.

within each group, systems can be compared using metrics specific to that approach. In our hypothetical example, in addition to meeting threat-based testing requirements, the soft systems might include a metric of conformity to body shape, while the hard systems might include a metric of specific hardness. Because our desire was not to have to insert weighting factors at this stage, in order to compare the groups on an equal basis, we also tried to define groups so that, to the extent possible, they had similar numbers of systems.

### Step 3—Estimate the Relative Values of Group Contributions

Having defined the groups representing alternative approaches to achieving the needed capability, we chose one approach on which to base our comparisons and defined this as the normative approach, or the "norm." This definition served only to provide a basis for comparison with the approaches of the other groups. We then assigned scores to each of the groups based on the following scale:

- 0—Makes no contribution to filling the capability gap.
- 1—Worse than the "norm" in filling the capability gap.
- 2—Same as the "norm" in filling the capability gap.
- 3—Better than the "norm" in filling the capability gap.
- 4—Much better than the "norm" in filling the capability gap.

The objective of this step was to provide a relative comparison of the approaches represented by the groups, i.e., these two approaches are about as good, but this one is better, and this one is much better. Thus, the choice of the "norm" was not important, but a critical review and intuitive evaluation of the resulting scores were.

We then disaggregated the total EV score proportionally among the groups. For example, if there were three groups that scored 2, 2, and 4, respectively, and the total gap EV were 0.5, this would be disaggregated as 0.125, 0.125, and 0.25, respectively.

### Step 4—Estimate the Relative Values of System Contributions to the Group EVs

To further disaggregate the group EVs into system EVs, we used the same scoring method as in Step 3 above. Here we had the advantage that we were comparing "apples to apples" as opposed to "apples to oranges," since each member of the group used the same approach to provide the needed capability. In our hypothetical example, we might have chosen systems based on Kevlar as our soft armor "norm" and systems based on silicon carbide as our hard armor "norm." As in Step 3, the choice of "norm" was less important than a critical review and intuitive evaluation of the resulting scores. We assigned scores to all of the systems in the group using the scale shown above and disaggregated the group EV proportionally among the systems in the group.

### Step 5—Estimate the Individual Contributions of Interdependent Systems

With Steps 1–4 completed, we had compared and scored relative to each other the alternative approaches to achieving the needed capabilities and the alternative systems that used each approach, and we had disaggregated the total EV accordingly. At this point, the FP capability gap total EV was fully disaggregated among its contributing systems, except that any interdependent systems identified in Step 1 still had a combined EV.

We used the following scheme to disaggregate the EV of the interdependent systems, which in practice would require subject matter expertise and could perhaps be accomplished via a Delphi exercise. We estimated the individual contributions of these interdependent systems by first asking which system was most important, i.e., which would on its own provide the greatest contribution. We then estimated how much greater this contribution would be than the contribution of all the other systems together, but without this one. For two interdependent systems, this immediately provided the individual contributions. For three or more interdependent systems, we repeated the process until the contributions became small enough that simply allocating the remainder evenly produced little difference.[6]

### Step 6—Perform a Sanity Check

The final step in our method for disaggregating the total EV for each FP capability gap was a two-part sanity check. The first part was simply to add the EVs of the individual systems and verify that the sum was in fact the total EV for the FP capability gap under consideration, which was either 0.25 or 0.50, as shown in Figure 3.3. The second part of this sanity check was a careful review of which systems received relatively high and low EVs to see if these assignments made sense in light of such factors as the nature of the gap, the nature of the system, the system's performance characteristics, and the relative EVs of competing systems.

## Distribution of System Expected Values

Following the procedures described above, we developed a matrix whose columns are the 22 non-red FP capability gaps, whose rows are the 183 contributing systems, and whose elements are the EV contributions of each system to each capability gap. This matrix is presented in the Appendix and used in Chapters Four and Five, together with estimates of remaining R&D cost and implementation cost, for portfolio optimization. Figure 3.4 shows the distribution of contributions of N/EMD-derived and RTBF systems to the 22 gaps, which we note varies significantly from gap to gap. This

[6] One can also treat the interdependency rigorously by using the constraints in the linear programming model. We did not do so because the performance data we used were not of adequate precision or accuracy.

**Figure 3.4**
**Distribution of N/EMD-Derived and RTBF System Contributions to Filling FP Capability Gaps**

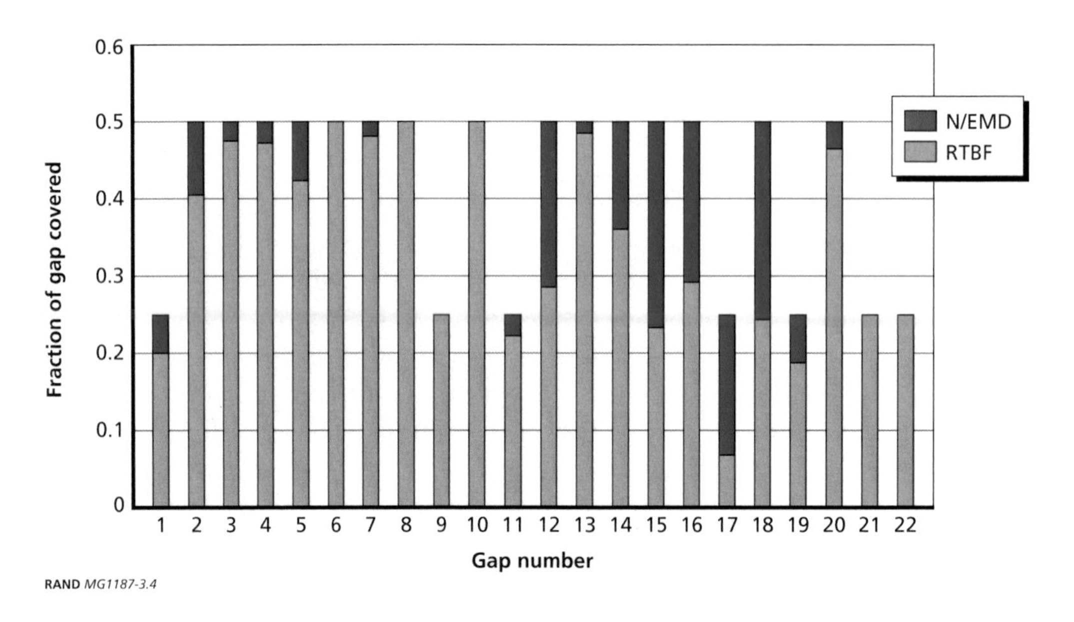


RAND *MG1187-3.4*

variation emphasizes the importance of portfolio analysis to identify which N/EMD projects need to be funded to continuation because systems derived from them make important contributions to filling specific capability gap requirements.

CHAPTER FOUR

# Optimal N/EMD Portfolio with Full Consideration of RTBF Systems

This chapter describes a demonstration of our PortMan method and model using the FP capability gap requirements,[1] N/EMD projects, and RTBF systems defined by TRADOC/ARCIC; the EV matrix described in Chapter Three and presented in the Appendix; and the remaining R&D cost and implementation cost estimates explained below. As stated previously, we distinguish between threshold (must have) and objective (desirable) requirements and use portfolio analysis to select an optimal portfolio of N/EMD projects based on analyses of the likelihood of meeting threshold requirements at different remaining R&D cost and implementation cost. We treat two different cases: (1) known budget and (2) uncertain budget.

In both cases, we assume that the Army has defined threshold requirements and objective requirements. For this demonstration, in which we have only the TRADOC/ARCIC-defined FP capability gap requirements, we take these capability gap requirements as the objectives and assume for this demonstration of our method that the threshold requirements are 75 percent of these objectives, as illustrated in Figure 4.1.[2]

## Remaining R&D and Implementation Cost Estimates

We obtained the remaining R&D costs of 19 of the 26 N/EMD-derived systems under consideration from the TRADOC/ARCIC quad charts described previously, which stated both the previous and projected R&D costs for each system depicted. For the remaining seven of the N/EMD-derived systems, the needed data were missing from these quad charts, and we derived the R&D cost data from the annual RDT&E Budget Item Justification Sheets (R-2 Exhibits). Figure 4.2 summarizes the remaining

---

[1] It should be emphasized that this study demonstrates the application of PortMan to FP capability gaps. Because FP is merely one of the ten gap areas, a real application would involve all ten areas.

[2] Some threshold requirements will be cheaper to meet than others. Our model can provide decision data on which requirements can be met and at what level when funds are limited.

**Figure 4.1**
**Threshold and Objective Requirements**

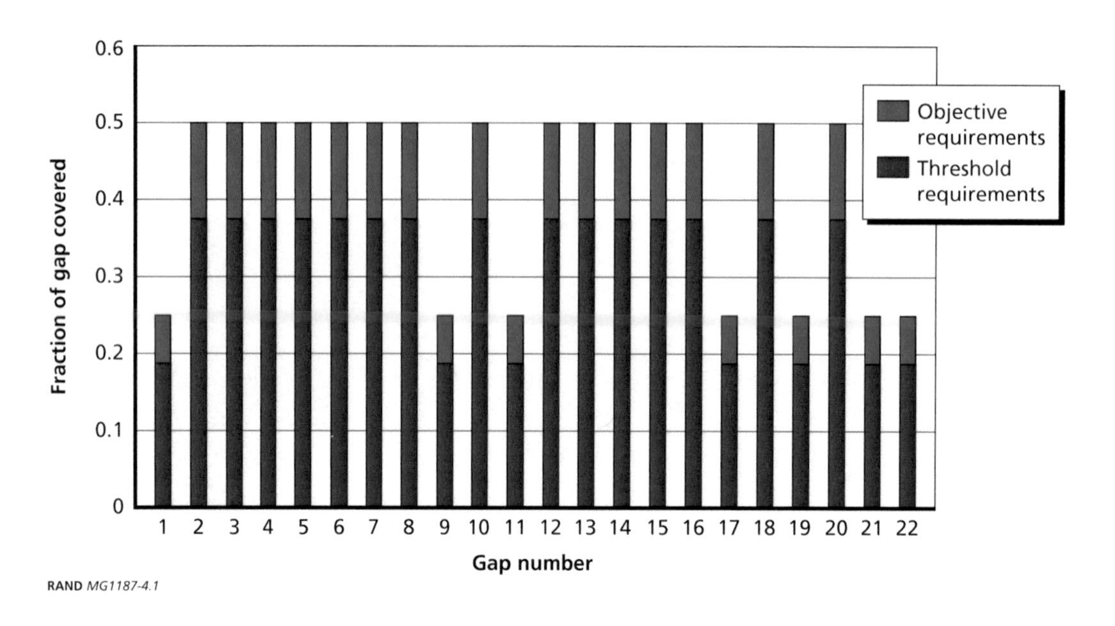


RAND *MG1187-4.1*

**Figure 4.2**
**Remaining R&D Cost Estimates for 26 N/EMD Systems**

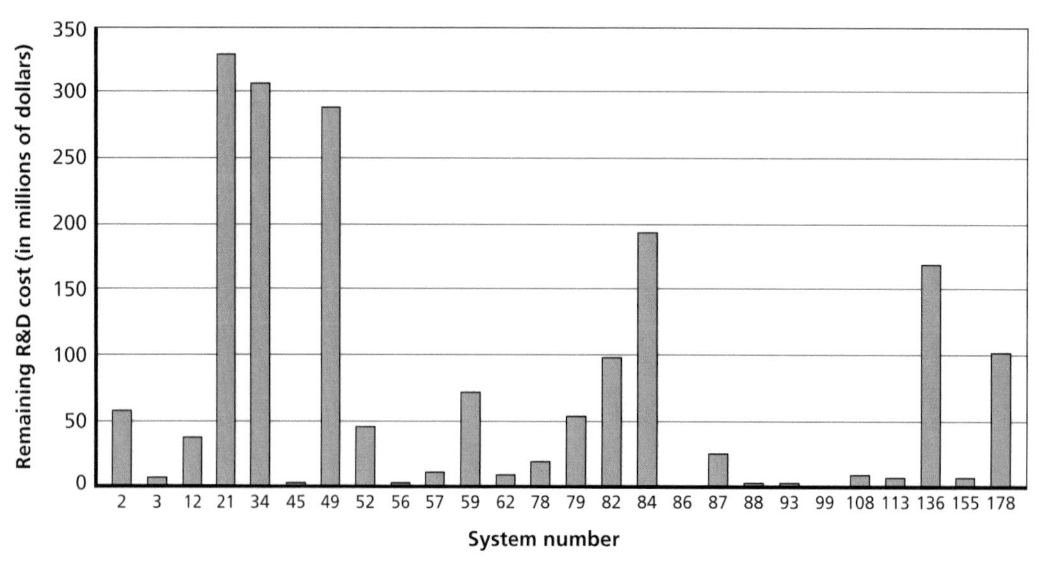


RAND *MG1187-4.2*

R&D cost estimates we obtained for the 26 N/EMD systems that were included in this demonstration.

Estimating implementation costs required considerably more effort than estimating remaining R&D costs. For example, the quad charts rarely provided basic data needed to estimate implementation cost, such as the expected unit cost of the system or the number of systems to be procured. Consequently, we made our implementation cost estimates using the same procedure used in the first two studies.[3] These implementation cost estimates were built up from estimates of unit system cost,[4] the number of systems to be acquired and fielded in order to fill the capability gaps, and the operating and maintenance cost of the systems over a 20-year planning horizon. Each of these cost estimates was for marginal cost[5] with respect to legacy systems; i.e., it was the additional estimated cost associated with replacing a legacy system with the new system under consideration. If the new system was estimated to be cheaper to acquire, operate, and maintain than the legacy system it was replacing, the implementation cost was negative.

To allow for uncertainty in these cost estimates, we defined the cost estimates made while the developing systems were in the N/EMD stage to be baseline implementation cost estimates. These baseline implementation cost estimates are shown in Figure 4.3. We then explicitly took into account uncertainty in these cost estimates by recognizing that the actual unit system cost when a system is fielded is often higher than that estimated when it was in the N/EMD stage. Consequently, we represented this uncertainty, for the purposes of our demonstration, by assigning a 0.5 probability that the eventual or actual cost of an N/EMD system is the same as the baseline cost estimate, and a 0.5 probability that the actual cost is twice the baseline cost estimate.[6] Taking into account this uncertainty, we estimate that the cost of RTBF systems is the average of the two equally probable cases—1.5 times the baseline cost estimate for the overrun case and 0.75 times the cost estimate for the savings case. Since we do not assume any uncertainty in the number of systems to be acquired and fielded,

---

[3] See Appendix C of TAS-1.

[4] The unit system cost is the acquisition cost of a system. As described in TAS-1, when such a cost is not provided, one can use the historic cost of a similar legacy system as the point of departure and adjust the cost by studying the difference between the system in question and the legacy system one chose for comparison.

[5] As noted earlier in the monograph, this study uses an at-the-margin approach to estimate costs and benefits of a new system with respect to the legacy system it replaces. The marginal cost is the cost above (or below, in the case of a savings) that of the legacy system. The marginal benefit is the benefit above and beyond what the legacy system can provide. Thus, the marginal cost used in this study is not the same as, and should not be confused with, the much more common usage of the term, which means the cost of the additional inputs needed to produce an additional unit of output.

[6] This assumption is for demonstration only. In a real application of the method, probabilities could, e.g., be based on experience with similar systems. If the baseline implementation cost estimate was negative, we assigned a 0.5 probability that the savings was half of that according to the baseline cost estimate, and another 0.5 probability that the savings was the baseline cost estimate.

**Figure 4.3**
**Baseline Implementation Cost Estimates for 183 Contributing Systems**

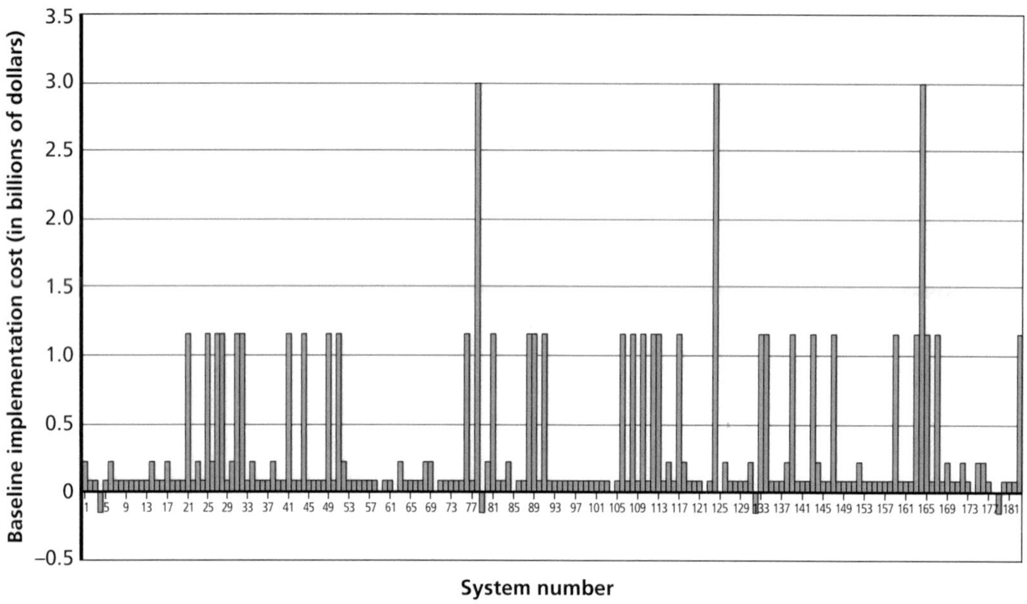

RAND *MG1187-4.3*

the expected implementation cost using these assumptions will vary from the baseline implementation cost by the same factor as the unit system cost.

## Optimal Portfolio for a Known Budget

This section describes our model demonstration for the first case mentioned in the previous chapter: a known budget. The input data to the model and simulation developed and demonstrated in TAS-1 and TAS-2 were the EV estimates for each of the N/EMD-derived and RTBF systems we consider, as shown in the Appendix, together with the R&D and implementation cost estimates shown in Figures 4.2 and 4.3. The model and simulation are described briefly below. More detailed descriptions can be found in the companion monographs. We then present and analyze the results of model and simulation runs aimed at maximizing the likelihood of meeting all threshold FP capability gap requirements shown in Figure 4.1 within the constraints of a given TRRD budget and a given TRLC budget.

### Linear Programming Model

The mathematical problem that the linear programming model solves is the selection of a subset of possible projects and systems that best meets a given objective under constraints such as those shown below. In TAS-1, the objective was to meet all requirements at the lowest TRLC cost for a given TRRD budget, without uncertainty.

For the current demonstration of the method, this can be represented mathematically as follows:

$$\text{Minimize} \sum_{i=1 \text{ to } 183} x_i \left( \text{RRDC}_i + \text{IC}_i \right)$$

subject to the constraint

$$\sum_{i=1 \text{ to } 183} x_i \left( \text{RRDC}_i \right) \leq \text{TRRD}$$

and a set of 22 constraints

$$\sum_{i=1 \text{ to } 183} x_i E\left[ V_{ij} \right] \geq TV_j \text{ for } j \text{ from 1 to 22 } \left( \text{FP capability gaps} \right),$$

where $x_i$ = 0 or 1 for nonselected and selected N/EMD projects and RTBF systems, respectively; $\text{RRDC}_i$ is the remaining R&D cost for the *i*th N/EMD project (zero for RTBF systems); $\text{IC}_i$ is the implementation cost for the *i*th N/EMD project or RTBF system; TRRD is the given total remaining R&D budget; $E[V_{ij}]$ is the EV contribution given in the Appendix for the *i*th (N/EMD-derived or RTBF) system to the *j*th FP capability gap; and $TV_j$ is the threshold value given in Figure 4.1 for the *j*th FP capability gap.

For the current model demonstration, similarly to that in TAS-2, the linear programming model is not used to find an optimal solution such as the minimum lifecycle cost in the problem described above, but rather to determine whether a solution exists that meets the given set of constraints. We then use the simulation described in the next section to determine the highest likelihood of meeting the set of 22 constraints shown above for any given TRRD cost and total remaining implementation cost.

### The Simulation

The simulation generates a set of outcomes by allowing each N/EMD-derived system's implementation cost to randomly take on either its estimated baseline value shown in Figure 4.3 or a value that is twice as high (a cost overrun).[7] Since there are 26 N/EMD projects under consideration, there will be numerous (i.e., 2 to the 26th power) sets

[7] See previous footnote for the negative implementation cost case. For the purposes of this demonstration, we assume that the implementation cost of RTBF systems is known because their EMD stage, where cost overruns typically occur, has already been completed. On the other hand, should one consider that even systems at the RTBF system stage would incur cost overruns when procured massively for fielding, one could include uncertainty in these implementation costs as well.

of outcomes. Each set of outcomes represents specifically which of the 26 projects will incur cost overruns in the unit cost of their systems and which will not. For example, a set might have cost overruns on systems derived from projects 1, 3, 4, 6, 9, 11, 12, 15, 23, 25, and 26, and no cost overruns on the rest of the 26 projects. We define the optimal portfolio as the subset of N/EMD projects that, if funded to completion, will provide systems that, together with RTBF systems, have the highest probability of meeting all threshold requirements (the set of 22 constraints described above) under a specific TRRD budget and a specific TRLC budget.[8]

Any given subset of N/EMD projects, or a portfolio, will produce many sets of outcomes in which different systems have cost overruns. For any specific set of outcomes, the key question is whether one can *select* from these N/EMD-derived systems with these cost characteristics (cost overruns or not) a subset that, together with RTBF systems, meets the set of 22 constraints shown above for a specific TRRD budget and for a specific TRLC budget. We are able to apply the linear programming model[9] to answer this question, since any specified set of outcomes has a specified implementation cost for each N/EMD-derived system.

Through analysis of characteristics of portfolios provided by the linear programming model under such specified certainty conditions, we have designed an algorithm to search for the optimal portfolio under uncertainty in implementation costs. Using this algorithm, we run the simulation 10,000 times for each portfolio. We define the feasible percentage as the percentage of these 10,000 runs for which the 22 constraints are met for given TRRD budget and TRLC budget. Then, the optimal portfolio is the portfolio that provides the highest possible feasible percentage.[10] The following section describes and analyzes the results of applying the linear programming model and simulation to find the optimal portfolio for a broad range of possible budgets.

### Description and Analysis of Results

Figure 4.4 shows the results of running the linear programming model and simulation for a known budget. Each point represents the highest possible feasible percentage for a given TRRD budget and TRLC budget[11] for all fielded systems, including both N/EMD-derived and RTBF systems. The TRLC cost for all 183 systems (if all were fielded) lies between $82 billion and $93 billion, depending on how many of the

---

[8] The TRLC budget is the sum of the TRRD budget and the total implementation budget.

[9] It should be noted that our linear programming model is typical and traditional in the sense that there is no uncertainty involved.

[10] In other words, the optimal portfolio is the subset of N/EMD projects that the model selects for funding and will have the highest probability of meeting the 22 requirements for given budgets. While for probability it is more appropriate to use *percent*, we use *percentage* instead because in our simulation approach it is the percentage of runs that are feasible.

[11] The TRLC budget is the sum of the TRRD budget and the total implementation budget. The uncertainty analyzed here is in the implementation budget for systems derived from N/EMD projects.

**Figure 4.4**
**Likelihood of Meeting Threshold Requirements Within a TRLC Budget for N/EMD-Derived and RTBF Systems**

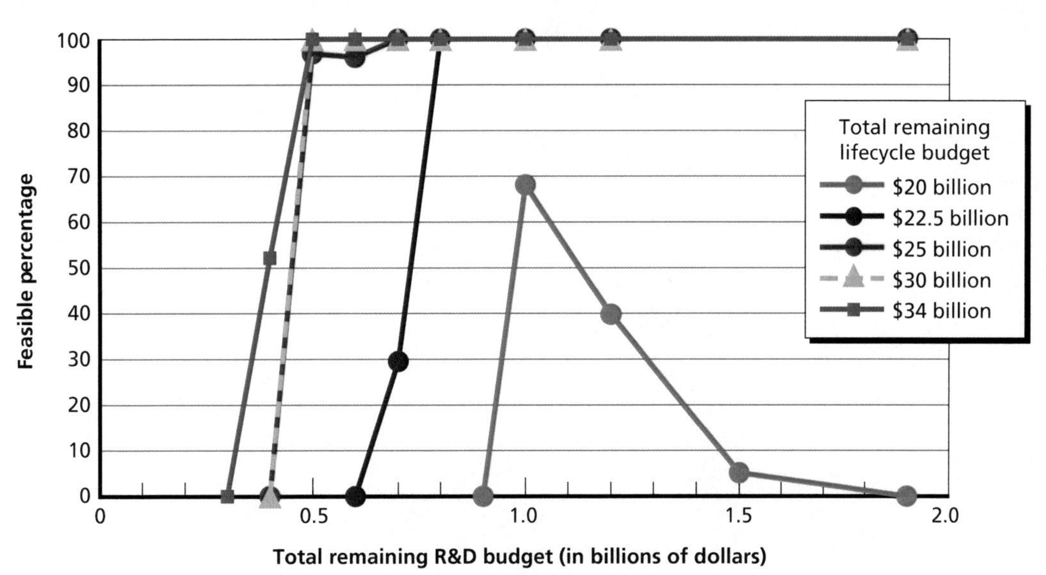


RAND *MG1187-4.4*

N/EMD-derived systems incur cost overruns. However, because of substantial redundancy in the capabilities of these systems, only a subset needs to be fielded to meet all threshold requirements.[12] In fact, our model runs show that TRLC budget in the range of \$30–34 billion provides a feasible percentage very close to 100 percent, but this percentage drops off sharply when the TRRD budget is reduced to below \$0.5 billion. Studying Figure 4.4, one sees that for each TRLC budget there is a point at which the feasible percentage falls off sharply, this point moves to higher TRRD budget as the TRLC budget decreases, and the maximum feasible percentage for lower TRLC budget is reduced substantially (e.g., to approximately 68 percent for a TRLC budget of \$20 billion).[13]

To find an optimal combination of TRLC budget and TRRD budget, we look for a "sweet spot" at which the feasible percentage (e.g., likelihood of meeting threshold requirements) is high and not too close to the rapid drop-offs described above. The

[12] The desired level of redundancy, e.g., to ensure that capabilities are sufficient to meet performance goals when outcomes are uncertain, is defined by the threshold requirement.

[13] We note that the seemingly counterintuitive decrease of feasible percentage with increase in TRRD budget for TRLC budget of \$20 billion (the green curve in Figure 4.4) results from the fact that for a fixed TRLC budget, increasing the TRRD budget requires decreasing the total implementation budget by the same amount. In this case, the decreased implementation budget is not enough to cover the cost of fielding all of the systems required to meet threshold requirements, especially considering the 50 percent probability of N/EMD-derived system cost overruns. Here the implementation, not the TRRD, budget is the binding constraint.

TRLC budget of $25 billion and TRRD budget of $0.7 billion present such a choice. The sweet spot is the most cost-effective budget with which most of the ongoing projects will be funded. The term also suggests that certain projects should be terminated, because the money saved from not funding them can be more cost-effectively spent on new projects.[14]

By presenting our results in the form of Figure 4.4, we emphasize the importance of considering lifecycle costs, effectively the "mortgage" to meet threshold requirements, when making R&D budget decisions. Since the lifecycle cost is the sum of R&D and implementation cost, this explicitly shows the trade-offs between the two. It is often the reality that R&D portfolio managers are provided with a given R&D budget, which will either be consumed or lost, with any residual funds not saved for implementation. We believe that the mindset leading to such choices should be altered, especially during the planning stage, and that Army leadership should instead optimally allocate funds between R&D and implementation. Such a strategy would help meet DoD's desire for savings from efficiency and effectiveness improvements. Further, the passing of the Budget Control Act on August 2, 2011, is a clear signal that the Army and the rest of DoD will continue to be active participants in the nation's efforts to function efficiently in an indefinitely austere budgetary environment. In this demonstration, our model results suggest a recommendation that the Army allocate $0.7 billion in TRRD budget to support the ongoing N/EMD projects and plan on needing $24.3 billion for implementation.

Table 4.1 shows that, for the sweet spot of a $25 billion TRLC budget and a $0.7 billion TRRD budget, the model selects 17 out of the 26 N/EMD projects for continued funding. These 17 projects form the optimal N/EMD portfolio that provides the greatest likelihood of meeting all threshold requirements—when the TRRD budget is $0.7 billion and the TRLC budget is $25 billion.

**Table 4.1**
**N/EMD Projects Selected for Continued Funding, with TRRD Budget of $0.7 Billion and TRLC Cost of $25 Billion for N/EMD-Derived and RTBF Systems**

| Project Number | 2 | 3 | 12 | 21 | 34 | 45 | 49 | 52 | 56 | 57 | 59 | 62 | 78 | 79 | 82 | 84 | 86 | 87 | 88 | 93 | 99 | 108 | 113 | 136 | 155 | 178 |
|---|---|---|---|---|---|---|---|---|---|---|---|---|---|---|---|---|---|---|---|---|---|---|---|---|---|---|
| Selected? | Y | Y | Y | N | N | Y | N | Y | Y | Y | Y | Y | N | N | Y | Y | Y | Y | N | N | Y | N | Y | N | Y | Y |

NOTE: Y = selected, and N = not selected.

[14] PortMan also identifies which requirements cannot be cost-effectively and fully met by the current N/EMD projects and suggests the targeted requirement gaps for the new projects to meet. Once the new projects have been so designed, PortMan can be used again to find the optimal portfolio from the current N/EMD and new projects, ensuring that the new projects are indeed cost-effectively designed to meet the gaps left behind by the current N/EMD projects. See TAS-2 for more information.

We note that the optimal portfolio shown in Table 4.1 is very different from that which would be selected using simpler, more intuitive, criteria. Figure 4.5 orders the N/EMD projects according to one such criterion that is often used for R&D project selection—the ratio of an N/EMD-derived system's total EV contribution to its remaining R&D cost in millions of dollars (total expected value [TEV]/remaining research and development [RRD] cost ratio). If the ordering shown in this figure were used for N/EMD project selection, the resulting portfolio would be quite different from our optimal portfolio. For example, project 84 would have been rejected according to TEV/RRD cost ratio because its low EV and/or high remaining R&D cost place it in the region of rejected projects (those in red). However, our model selects it because, although its cost-effectiveness is very low, it is needed to meet some individual threshold requirements. On the other hand, projects 88 and 108, with attractive cost-effectiveness and located among selected projects (those in green), would have been selected according to TEV/RRD cost ratio but are rejected by our model because the threshold requirements to which they contribute can be met by even more cost-effective projects.

The reason for the differences between the results in Table 4.1 and those shown in Figure 4.5 is that the linear programming model and simulation take into account the interplay between requirements and N/EMD-derived and RTBF systems in a way that no single criterion can reproduce. For example, a system typically contributes to

**Figure 4.5**
**Comparison of N/EMD Project Ordering According to TEV/RRD Cost Ratio with Optimal Portfolio Selections**

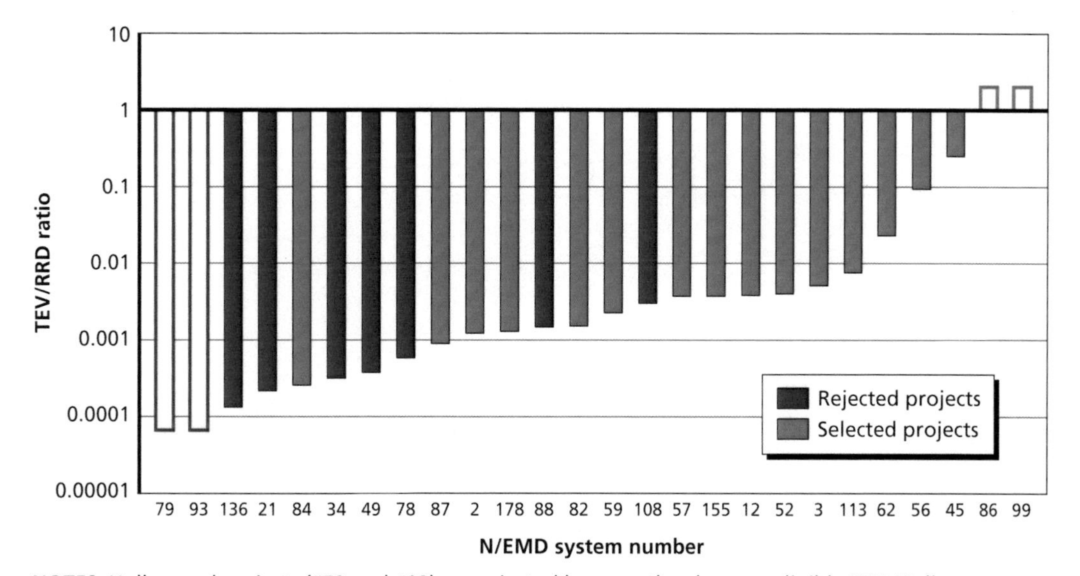


NOTES: Hollow-red projects (#79 and #93) are rejected because they have negligible TEV. Hollow-green projects (#86 and #99) are selected because they have negligible RRD cost.
RAND *MG1187-4.5*

multiple requirements and a requirement is generally met by multiple systems. Further, the selection of N/EMD projects for continued funding must consider capabilities that could otherwise be supplied by RTBF systems. No simple criterion is likely to be able to account for the interaction of these complex factors, a task for which our model was specifically designed.

## An Optimal Portfolio for an Uncertain Budget

The TRLC budget is often uncertain at the time that a decision to continue or terminate N/EMD projects must be made. To address such situations, we suggest selecting the N/EMD portfolio under the constraint of maximizing feasible percentage, with a defined set of weights assigned to future plausible budgets. To illustrate this approach, we assume in our demonstration that the TRLC budget is equally likely to be $20 billion, $22.5 billion, or $25 billion.[15] Figure 4.6 shows the results of running our model under these conditions. These results suggest that under this budget uncertainty, it is best to provide $1 billion for the remaining R&D of a selected subset of the ongoing N/EMD projects, which are shown in Table 4.2. Figure 4.6 indicates that this selected portfolio will have an 83 percent likelihood of meeting all threshold requirements

**Figure 4.6**
**Likelihood of Meeting Threshold Requirements, with TRLC Budget Equally Likely to Be $20 Billion, $22.5 Billion, or $25 Billion**

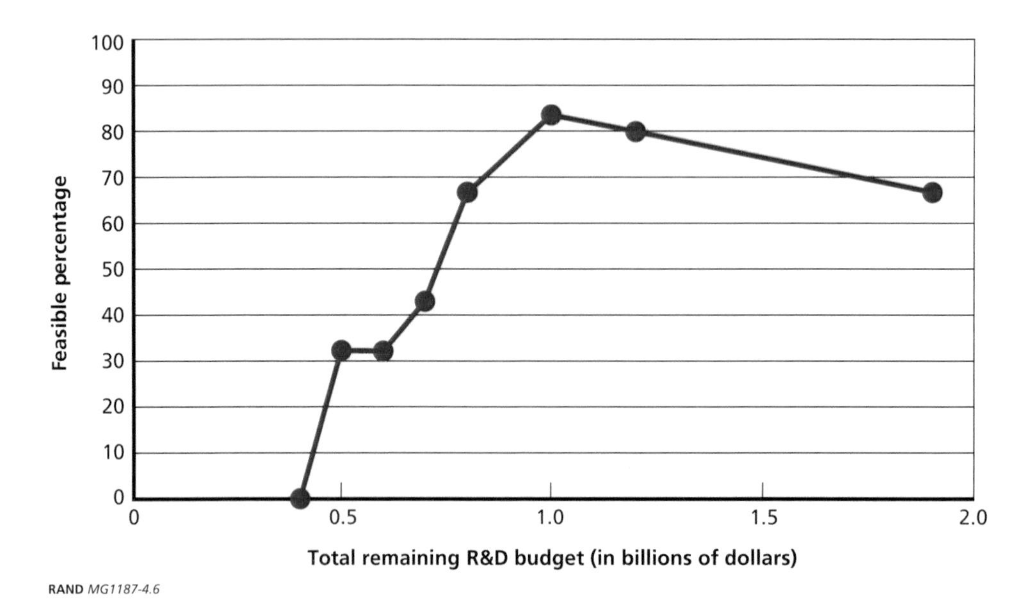


[15] We assume that the TRLC budget uncertainty is in the total implementation budget, not the TRRD budget.

**Table 4.2**
**N/EMD Projects Selected for Continued Funding, with TRRD Budget of $1 Billion and TRLC Budget for N/EMD-Derived and RTBF Systems with Equal Probability of $20 Billion, $22.5 Billon, or $25 Billion**

| Project Number | 2 | 3 | 12 | 21 | 34 | 45 | 49 | 52 | 56 | 57 | 59 | 62 | 78 | 79 | 82 | 84 | 86 | 87 | 88 | 93 | 99 | 108 | 113 | 136 | 155 | 178 |
|---|---|---|---|---|---|---|---|---|---|---|---|---|---|---|---|---|---|---|---|---|---|---|---|---|---|---|
| Selected? | Y | Y | Y | N | Y | Y | N | Y | Y | Y | Y | Y | N | N | Y | Y | Y | Y | N | N | Y | Y | Y | N | Y | Y |

NOTE: Y = selected, and N = not selected.

with the specified uncertainty in TRLC budget, i.e., equal probability of $20 billion, $22.5 billion, or $30 billion.[16] It also indicates that the feasible percentage drops off rapidly from this value for TRRD budgets less than $1 billion. Thus, the model not only finds the optimal portfolio under budget uncertainty but also identifies a "threshold" TRRD budget for the Army to stay above to maintain a reasonable chance of meeting its threshold requirements.

For a TRLC budget that is equally likely to be $20, $22.5, and $25 billion and a TRRD budget of $1 billion, the model suggests terminating the seven N/EMD projects shown in Table 4.2. We compare this to Table 4.1, which is the optimal portfolio for a known TRLC budget of $25 billion and TRRD budget of $0.7 billion. For the certainty case shown in Table 4.1, the model suggests terminating nine N/EMD projects. While all seven N/EMD projects rejected in the uncertainty case are also rejected in the certainty case, the latter rejected two more, projects 34 and 108. This is not surprising, because in the face of an uncertain budget, some projects are retained as a hedge against uncertainty. This also shows that the difference in optimal portfolio when uncertainty is taken into account can include differences in project acceptance and project rejection. Ignoring uncertainty in the TRLC budget may thus lead to a suboptimal selection of ongoing N/EMD projects for continued funding.

## Comparison of PortMan and Other Models with Full Consideration of RTBF Systems

It is of interest to compare the results of PortMan with those from other models. Generally, models can be classified into three groups. Models in the first group are based on individual projects' abilities to meet requirements at low costs but do not consider the synergistic (i.e., portfolio) effects among projects that the other two groups do. A model based on a ratio of the TEV over the RRD cost belongs to this group. It would first pick the N/EMD project with the highest TEV to RRD cost ratio (i.e., the project

[16] The model can handle any user-specified distribution of uncertainties in the TRLC budget.

on the far right in Figure 4.5). It would then pick the project with the next highest ratio and so on until the TRRD budget was fully committed. The feasible percentages are shown under the column Benefit/Cost Ratio Model in Table 4.3.

The second group of models considers not only the merits of the individual projects but also such portfolio effects as their combined abilities to meet multiple requirements. While the first group could select two projects having the highest benefit/cost ratios, both projects might actually be credited with high benefits for being able to meet requirement 1 extremely well. As a consequence, they could be credited with a high overall benefit score in spite of the fact that they could only meet requirement 2 poorly or not at all. On the other hand, models in the second group would always select a portfolio of projects that can meet all requirements, even if only barely, over one that produces values greatly above most requirements but barely fails to meet the other few requirements. Still, the second group does not account for uncertainties and, thus, is optimization under certainty. When there are two estimates of a system's cost, a group-2 model would use a single cost, namely, the expected cost. The resulting feasible percentages are shown under the Certainty Model column in Table 4.3.

Models in group 3 are the same as those in group 2, except for one major difference—in group 3's ability to account for uncertainties. PortMan belongs to this group. PortMan's linear programming model and simulation are able to weigh simultaneously the complex interplay of requirements, system capabilities, costs, and uncertainties in a way that no simple criterion or set of criteria in the other two groups can reproduce.

Table 4.3 is a comparison of model results with full consideration of RTBF systems. We compare the results for cases with different combinations of TRRD budget and TRLC budget. At the reference case of $0.7 billion in TRRD budget and $25 billion in TRLC budget as shown in Figure 4.4, PortMan's optimal portfolio meets all requirements practically for certain (i.e., feasible percentage to be 100 percent). This

**Table 4.3**
**Comparison of PortMan and Other Models with Full Consideration of RTBF Systems**

| Budget | | Feasible Percentage | | |
|---|---|---|---|---|
| TRRD (in billions of dollars) | TRLC (in billions of dollars) | PortMan | Certainty Model | Benefit/Cost Ratio Model |
| 0.7 | 25 | 100% | 100% | 28% |
| 1.0 | 20 | 68% | 56% | 0% |
| 1.2 | 20 | 40% | 0% | 0% |

budget combination has been selected as the sweet spot or the reference case because the optimal portfolio has a built-in safety margin that keeps the feasible percentage high even if the TRRD budget turns out to be lower.[17] This large safety margin allows the optimal portfolio derived from the certainty model to attain 100 percent feasibility as well.[18] In contrast, the benefit/cost ratio model performs poorly to produce an "optimal" portfolio whose feasible percentage is merely 28 percent, far below the 100 percent attained by the certainty model and PortMan.

For the case of $1.0 billion for TRRD and $20 billion for TRLC as shown in Figure 4.4, PortMan's portfolio has a feasible percentage of 68 percent, while the certainty model gives 56 percent and the benefit/cost ratio model 0 percent. The certainty model does not fare as well because it has not accounted for the uncertainties. Project selection based on the benefit/cost ratio model simply cannot meet all requirements, because its selection is based on the sum of a project's contributions to requirements and does not consider the portfolio effects.

For the case of $1.2 billion for TRRD and $20 billion for TRLC as shown in Figure 4.4, PortMan yields 40 percent, while the other two models have little chance to meet all requirements.

In sum, for a future inevitably full of uncertainties, a model such as PortMan that accounts for uncertainties can select the optimal portfolio with the highest chance to meet all requirements. In some specific cases, a certainty model fares as well as PortMan, but in other cases, the certainty model can be much worse. Further, one cannot tell, a priori, when the certainty model will fare as well as, or close to, PortMan. In all cases, the traditional benefit/cost ratio model fares worse than even the certainty model, clearly showing the importance of considering the portfolio effect.

---

[17] The feasible percentage remains at 100 percent if the TRRD budget drops from $0.7 billion to $0.6 billion. Even if it drops to $0.5 billion, the feasible percentage is high, 96 percent.

[18] On the other hand, a certainty model does not necessarily attain the same feasibility as PortMan at the sweet spot. See the first row in Table 5.3. Moreover, for the sweet spot in TAS-2 (Figure S.5, p. xxii), while PortMan attains 91 percent feasibility, a certainty model reaches only 45 percent.

CHAPTER FIVE

# An Optimal N/EMD Portfolio, with Approximate Consideration of RTBF Systems

Chapter Four describes a procedure to arrive at an optimal portfolio of N/EMD projects with full consideration of RTBF systems. Since the cost and performance data for RTBF systems are sometimes incomplete or inadequately updated, we have developed a portfolio analysis method with approximate consideration of RTBF systems.

The first step in this approximate consideration method is to assemble the contributions of the N/EMD-derived systems to the capability gaps identified by TRADOC/ARCIC. Figure 5.1 shows the total EV contributions of these 26 systems to each of the 22 capabilities, using the EVs in the Appendix.[1]

**Figure 5.1**
**Contributions of 26 N/EMD-Derived Systems to Requirements**

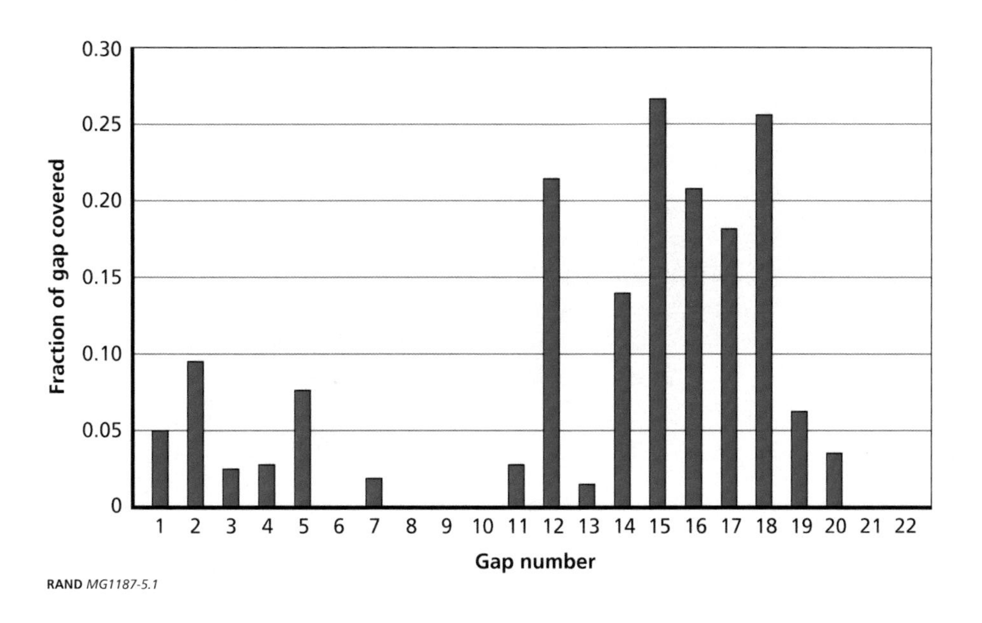


[1] The total EV contributions of these 26 N/EMD systems are also shown in Figure 3.4.

The second step is to determine whether any of the N/EMD-derived system contributions can be made more cost-effectively by RTBF systems instead. Unlike in Chapter Four, we assume in this chapter that we do not have sufficient data to make this determination. For this demonstration only, we assume that all of the N/EMD projects are providing capabilities that could not be provided by RTBF systems.

The third step is to determine if there are projects that have not yet reached the N/EMD stage but may contribute to capability gaps for which immediate filling is not essential. Again, we do not have sufficient data to make this determination, but the focus of this chapter is not on earlier-stage R&D versus N/EMD projects, but rather on finding out what happens when RTBF systems are not adequately considered. So, for this demonstration only, we assume that the Army cannot afford to wait for future N/EMD projects to fill the gaps that the ongoing N/EMD projects are addressing.

The simplifying assumptions of the previous two steps are made only to demonstrate our approximate consideration method. Given adequate data for analysis, these two steps can allow adjustments in the N/EMD project portfolio based on the characteristics of relevant RTBF systems and future N/EMD projects. However, assumptions as unrealistic as those made above are sometimes used in actual cases. Comparison of the simplistic optimal portfolio in this chapter and the more realistic optimal portfolio shown in Chapter Four indicates the level of sub-optimality that may ensue from such simplistic assumptions.

The fourth step is to determine the threshold (minimum) requirements for the N/EMD-derived systems to meet. We assume that the objective (desirable) requirements are the TEVs provided by all 26 N/EMD-derived systems. While the differences between individual threshold requirements and their corresponding objective requirements are likely to be variable, we assume for this demonstration that each threshold requirement is 75 percent of its objective requirement.[2] Figure 5.2 shows these threshold and objective requirements.[3]

## An Optimal Portfolio for a Known Budget

Figure 5.3 shows the results of running the linear programming model and simulation for the case in which the budget is known. Each point represents the highest possible feasible percentage for a given TRRD budget for the selected N/EMD projects and TRLC budget for all systems fielded from them. The TRLC cost for all 26 N/EMD-derived systems (if all were fielded) lies between \$10 billion and \$21 billion, depending

[2] Our model can use any user-defined threshold. Further, the user can vary the threshold level in order to see the costs of meeting various levels and to select a level that can be met cost-effectively.

[3] We note that these requirements, which are based on only the blue portions of Figure 3.4, are much less demanding than the requirements used in Chapter Four, when fully considering RTBF systems.

**Figure 5.2**
**Objective and Threshold Requirements for the 26 N/EMD-Derived Systems**

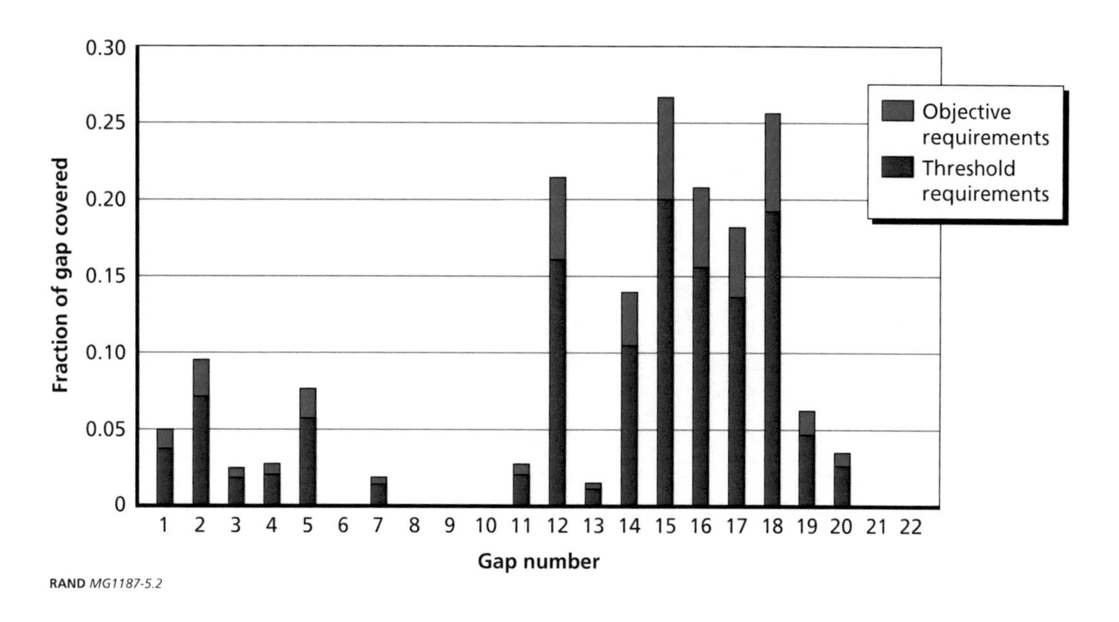


**Figure 5.3**
**Likelihood of Meeting Threshold Requirements Within a TRLC Budget for N/EMD-Derived Systems**

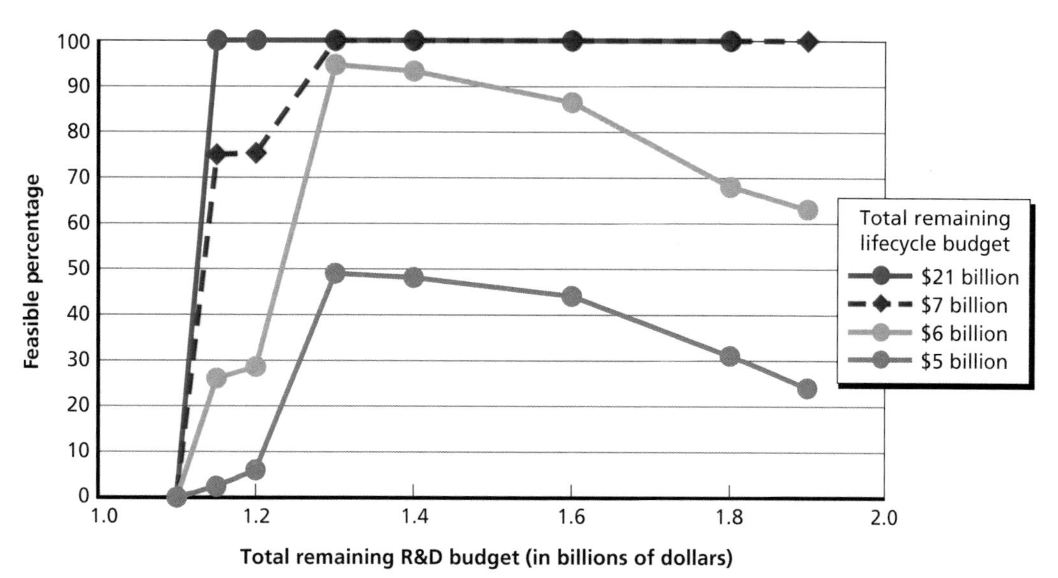

on how many of the systems incur cost overruns. However, as discussed in Chapter Four with respect to Figure 4.4, because of substantial redundancy in the capabilities of these systems, only a subset needs to be fielded to meet all threshold requirements. In fact, Figure 5.3 shows that a TRLC budget of $7 billion provides very close to 100 feasible percentage, as long as the TRRD budget does not fall below $1.3 billion. Examination of the variation of feasible percentage with TRLC budget and TRRD budget suggests a sweet spot, or point of best cost-effectiveness, near $7 billion for the TRLC budget and $1.4 billion for the TRRD budget, as $1.3 billion would have been too close to the cliff and have too little a safety margin in the event that the cost estimates are only somewhat off.

For a TRLC budget of $7 billion and TRRD budget of $1.4 billion, the model selects 19 N/EMD projects for continued funding and suggests terminating the following 7 projects: 45, 49, 78, 88, 93, 108, and 136, as shown in Table 5.1.

We now compare the rejected projects with those rejected for the known budget case in Chapter Four, where full consideration was given to RTBF systems.

- For the known budget case in Chapter Four, with TRRD budget for N/EMD projects of $0.7 billion and TRLC budget for N/EMD-derived and RTBF systems of $25 billion, the rejected N/EMD projects were 21, 34, *49, 78*, 79, *88, 93, 108,* and *136*, as was shown in Table 4.1.
- As shown in Table 5.1, with TRRD budget for N/EMD projects of $1.4 billion and TRLC budget for N/EMD-derived systems of $7 billion, the rejected N/EMD projects are 45, *49, 78, 88, 93, 108,* and *136.*

This comparison reveals two important findings. First, failure to fully consider the RTBF systems results in a much higher TRRD budget for N/EMD projects—$1.4 billion versus $0.7 billion.[4] This underscores the importance of considering already avail-

**Table 5.1**
**N/EMD Projects Selected for Continued Funding with TRRD Budget of $1.4 Billion and TRLC Budget of $7 Billion for N/EMD-Derived Systems**

| Project Number | 2 | 3 | 12 | 21 | 34 | 45 | 49 | 52 | 56 | 57 | 59 | 62 | 78 | 79 | 82 | 84 | 86 | 87 | 88 | 93 | 99 | 108 | 113 | 136 | 155 | 178 |
|---|---|---|---|---|---|---|---|---|---|---|---|---|---|---|---|---|---|---|---|---|---|---|---|---|---|---|
| Selected? | Y | Y | Y | Y | Y | N | N | Y | Y | Y | Y | Y | N | Y | Y | Y | Y | Y | N | N | Y | N | Y | N | Y | Y |

NOTES: Y = selected, and N = not selected. While project 79 has a zero total EV, it has a negative (marginal) implementation cost or a cost savings if it, instead of the more expensive legacy system it replaces, is procured and fielded. For this case, it is selected because it helps reduce total implementation cost.

[4] The much higher TRLC budget of $25 billion for the known budget case in Chapter Four reflects the much more demanding requirements when RTBF systems are fully considered.

able capabilities before investing in R&D aimed at developing new systems. Second, there are six N/EMD projects[5] (italicized in the bullets above) that are rejected whether RTBF systems are fully considered or not. These projects should be the first ones terminated under budget constraints. Investments in new N/EMD projects will likely be superior to continuing such projects.

Figure 5.4 shows how the N/EMD projects selected by our model differ from those that would be selected based on ranking according to the ratio of total EV contribution to remaining R&D cost. In particular, several of the rejected projects in the italicized group above, which were also rejected when fully considering RTBF systems, would be accepted using the simple ratio rule. Our model rejects project 45. Even though it has a very attractive ratio, we find other projects, in particular Project 86, are even more cost-effective, making the inclusion of project 45 an inferior choice.[6] Again,

**Figure 5.4**
**Comparison of N/EMD Project Ordering According to TEV/RRD Cost Ratio with Optimal Portfolio Selections**

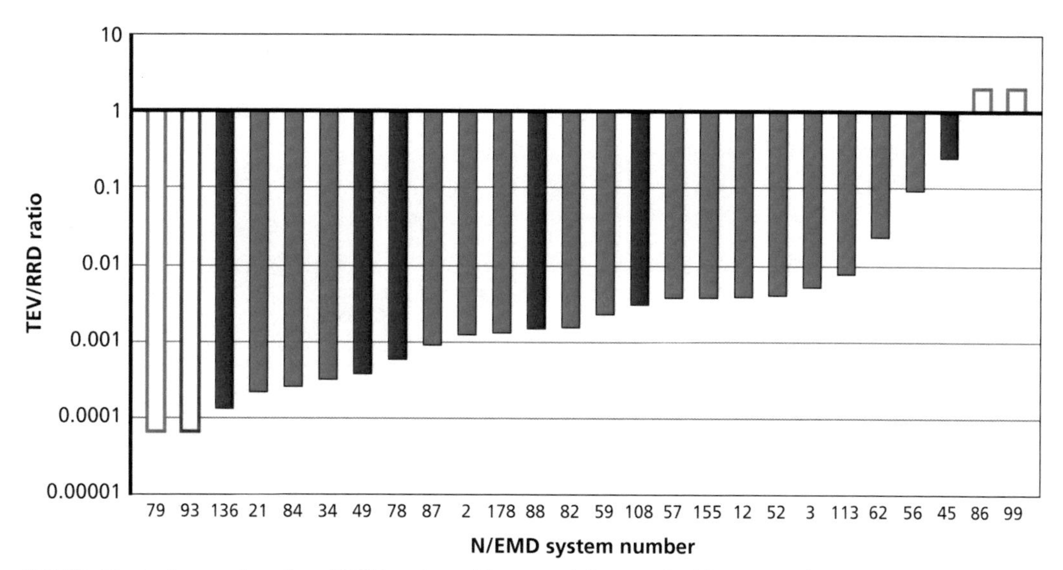


NOTES: The hollow-red project (#93) is rejected because it has negligible TEV. Hollow-green projects are selected because either they (#86 and #99) have negligible RRD cost or it (#79) has a negative implementation cost.
RAND *MG1187-5.4*

---

[5] Project 45 is rejected when RTBF systems are approximately considered, but it is kept when RTBF systems are fully considered.

[6] As was shown in Figure 4.5, project 45 is selected when RTBF systems are fully considered. This demonstrates the difficulty of reproducing the most cost-effective portfolio through approximate considerations of RTBF systems.

as in Figure 4.5, the model takes into account the intricate interactions among many factors when selecting the project portfolio.

## An Optimal Portfolio for an Uncertain Budget

Figure 5.5 shows the results of running the linear programming model and simulation for the case in which the budget is uncertain. More specifically, it shows the maximum feasible percentage as a function of TRRD budget for N/EMD-derived systems, with our approximate treatment of RTBF systems and assuming equal probability that the TRLC budget is \$5 billion, \$6 billion, or \$7 billion. In this case, the maximum feasible percentage of 80 percent occurs for a TRRD budget of \$1.4 billion. The model results also show that the maximum feasible percentage for this uncertain budget with equal probability of \$5 billion, \$6 billion, or \$7 billion drops off rapidly for a TRRD of less than \$1.4 billion. This behavior is similar to that shown in Figure 4.6 for the uncertain budget case with full consideration of RTBF systems. It suggests that, for maximum cost-effectiveness in this demonstration case, with the assumptions described above, the TRRD budget should be kept above \$1.4 billion.

Table 5.2 shows the N/EMD projects selected for continued funding for the most cost-effective TRRD of \$1.4 billion. In this case, the model selects 19 projects and recommends terminating seven projects. Comparison of Tables 5.1 and 5.2 shows that

**Figure 5.5**
**Likelihood of Meeting Threshold Requirements Within a TRLC Budget for N/EMD-Derived Systems Equally Likely to Be \$5 Billion, \$6 Billion, or \$7 Billion**

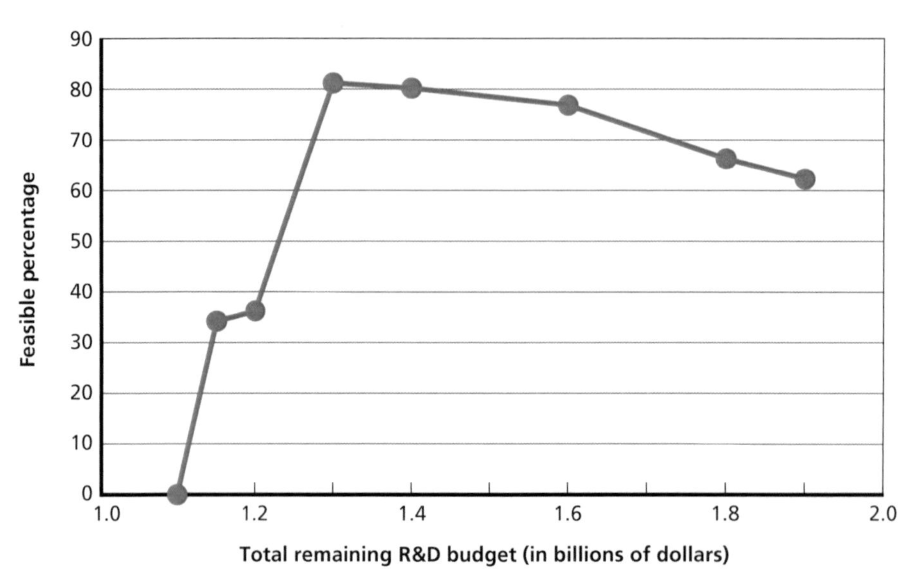


RAND *MG1187-5.5*

**Table 5.2**
**N/EMD Projects Selected for Continued Funding with TRRD Budget of $1.4 Billion and TRLC Budget for N/EMD-Derived Systems with Equal Probability of $5 Billion, $6 Billion, and $7 Billion**

| Project Number | 2 | 3 | 12 | 21 | 34 | 45 | 49 | 52 | 56 | 57 | 59 | 62 | 78 | 79 | 82 | 84 | 86 | 87 | 88 | 93 | 99 | 108 | 113 | 136 | 155 | 178 |
|---|---|---|---|---|---|---|---|---|---|---|---|---|---|---|---|---|---|---|---|---|---|---|---|---|---|---|
| Selected? | Y | Y | Y | Y | Y | N | N | Y | Y | Y | Y | Y | N | Y | Y | Y | Y | Y | N | N | Y | N | Y | N | Y | Y |

NOTE: Y = selected, and N = not selected.

project selection for TRRD of $1.4 billion is identical whether the TRLC budget is known or uncertain. However, this result is fortuitous; as shown in Tables 4.1 and 4.2, there are differences in N/EMD project selection for known and uncertain budgets when RTBF systems are fully considered.

## Comparison of PortMan and Other Models with Approximate Consideration of RTBF Systems

This section is the same as the one in Chapter Four, except the consideration of RTBF systems is approximate, not full. Since PortMan considers both portfolio effects and uncertainties, its optimal portfolio will have a higher likelihood to meet all requirements for any given TRRD and TRLC budgets. The certainty model fares worse in all cases, and the extent of sub-optimality is case-dependent. Table 5.3 shows that the use of a certainty model would lead to mild sub-optimality in three cases (77 percent instead of 86 percent, 60 percent instead of 68 percent, and 56 percent instead of 63 percent). The certainty model would lead to a moderately sub-optimal case (74 percent instead of 100 percent) and two severely sub-optimal cases (25 percent instead of 95 percent and 25 percent instead of 93 percent). Worse yet, one cannot use the certainty model only in the mildly sub-optimal cases because, without PortMan, one cannot tell in advance which cases would yield only mild sub-optimality.

For the first four cases in Table 5.3, project selection based on benefit-to-cost ratios would simply fail to meet all threshold requirements. While occasionally this model can yield somewhat higher feasible percentages than the certainty model, as in the last two cases in Table 5.3, one cannot tell in advance the types of cases for which the benefit/cost ratio model is better.

**Table 5.3**
**Comparison of PortMan and Other Models with Approximate Consideration of RTBF Systems**

| Budget | | Feasible Percentage | | |
|---|---|---|---|---|
| TRRD (in billions of dollars) | TRLC (in billions of dollars) | PortMan | Certainty Model | Benefit/Cost Ratio Model |
| 1.4 | 7 | 100% | 74% | 0% |
| 1.3 | 6 | 95% | 25% | 0% |
| 1.4 | 6 | 93% | 25% | 0% |
| 1.6 | 6 | 86% | 77% | 0% |
| 1.8 | 6 | 68% | 60% | 61% |
| 1.9 | 6 | 63% | 56% | 63% |

CHAPTER SIX

# Findings and Recommendations

During this study, we applied our PortMan method and model to selecting portfolios of R&D projects in the EMD stage of the Defense Acquisition and Management System. To identify the most cost-effective investments in these EMD projects, we also considered capabilities that could be obtained by procuring and fielding RTBF systems instead. For capabilities for which fielding could be delayed, we considered as well (NEMD) projects that were near, but not yet ready, to enter the EMD stage.

In TAS-2, we demonstrated that an optimal portfolio of ATOs (the Army's highest priority R&D projects) that met all capability gap requirements when uncertainty in project success was neglected had a very small (about 16 percent) likelihood of meeting these same requirements when such uncertainty was taken into account. This demonstration of the necessity of accounting for uncertainty when selecting project portfolios led us to include in our analysis of N/EMD portfolios uncertainties in the cost of N/EMD-derived systems (e.g., from cost overruns) and in the budget available to cover the lifecycle costs of N/EMD-derived and RTBF systems.

Because cost and performance data for RTBF systems are often incomplete or inadequately updated, in addition to running PortMan with full consideration of RTBF systems, we developed a method that allowed us to select N/EMD portfolios with approximate consideration of RTBF systems. We then applied both methods to the cases of a known and an uncertain TRLC budget for N/EMD-derived and RTBF systems.

## Findings

Based on the results and analyses shown and described in previous chapters, we report the following findings:

- We have demonstrated the use of the PortMan method and model for N/EMD portfolios to
  - find the sweet spot of a cost-effective TRRD budget for N/EMD projects and TRLC budget of N/EMD-derived and RTBF systems that provides a high

  likelihood of meeting requirements when the system costs, and consequently the implementation costs, are uncertain
  - select the optimal portfolio of N/EMD projects for funding, with TRRD budget and TRLC budget at the sweet spot
  - select the portfolio of N/EMD projects that provides the highest likelihood of meeting requirements for any given TRRD budget and TRLC budget.
- Similar to results shown in TAS-2 for ATOs, the optimal portfolio of N/EMD projects is not the same as would result from the use of simple criteria such as the ratio of TEV to TRRD cost.
- Taking into account uncertainty in the budget for TRLC cost (specifically the implementation budget) of N/EMD-derived and RTBF systems makes a difference for both projects selected and projects rejected in N/EMD portfolios, signifying that, if one ignores the inevitable uncertainties, one would end up selecting a sub-optimal portfolio.
- Taking into account uncertainty in the TRLC budget allows identification of a threshold ("must-have") TRRD budget below which the likelihood of meeting requirements decreases sharply, alerting Army planners to the dire consequences of a budget below that threshold.
- A TRRD budget for the optimal N/EMD portfolio is significantly less when RTBF systems are fully considered, allowing the Army to take best advantage of the many systems that have already been developed.
- Comparison of PortMan with full and with approximate consideration of RTBF systems allows the identification of N/EMD projects (rejected in both portfolios) that do not make important contributions to requirements and are, thus, prime candidates for termination.

## Recommendations

Based on these findings, we make the following recommendations, which are based on the results described in Chapters Four and Five:

- When selecting portfolios of EMD projects, the Army should evaluate the performance and cost trade-offs of obtaining the same or similar capabilities by fielding RTBF systems or, when fielding can be delayed, by developing and fielding N/EMD-derived systems. This evaluation should consider contributions to meeting requirements, when these contributions can be achieved, at what cost, and with what risk.
- When comparing N/EMD-derived and RTBF systems, the Army should take into account uncertainties in the cost of N/EMD-derived systems and in the budget for TRLC cost of N/EMD-derived and RTBF systems.

- When selecting N/EMD project portfolios, the Army should seek to balance TRRD budget and TRLC budget[1] for systems derived from selected N/EMD projects and RTBF systems in order to maximize the likelihood of meeting requirements. Rather than allocating R&D funds separately on a use-it-or-lose-it basis, we recommend that, especially at the planning stage, Army leaders should optimally allocate lifecycle budget between R&D and implementation, applying funds saved in R&D toward the implementation cost of systems derived from successful R&D projects. In Chapter Four, we demonstrate how to identify such an optimal allocation. This change in mindset could help the Army meet DoD objectives for savings from efficiency and effectiveness, which are particularly critical as the nation is in an austere budgetary environment after the passing of the Budget Control Act on August 2, 2011.
- The Army should evaluate a wide enough range of plausible future lifecycle cost budgets for N/EMD-derived and RTBF systems to identify threshold RRD budgets for N/EMD projects, below which the likelihood of meeting requirements decreases rapidly. In Chapters Four and Five, we illustrate this using a 20 percent to 30 percent range of total lifecycle cost budget for N/EMD-derived and RTBF systems.

In conclusion, we note that it is important in portfolio analysis not only to consider lifecycle costs and uncertainties in performance, cost, and budget but also to bring these factors into the evaluation on a consistent basis. The PortMan method and model demonstrated under this study and described in this monograph provide a means for the Army to accomplish both of these ends.

---

[1] A TRLC budget equals a TRRD budget plus total implementation budget.

APPENDIX

# Expected Values of N/EMD-Derived and RTBF Systems

This Appendix presents the matrix of EV contributions of the 183 systems to the 22 FP capability gaps considered in this monograph. The values in the matrix were estimated using the method described in Chapter Three and were used as inputs to the linear programming model and simulation to produce the results presented and analyzed in Chapters Four and Five. In Tables A.1–A.11, red boxes denote that the system makes no contribution to the gap. Yellow boxes mean that the system was not included in the disaggregation of total expected value because of a lack of necessary data. Green boxes denote that the system makes a contribution to the gap, and the expected value contribution of that system to that gap is shown.

**Table A.1**
**Expected Values of Systems 1 to 17**

| System Name | Project Number | Developmental Stage | Gap Number | | | | | | | | | | | | | | | | | | | | | |
|---|---|---|---|---|---|---|---|---|---|---|---|---|---|---|---|---|---|---|---|---|---|---|---|---|
| | | | 1 | 2 | 3 | 4 | 5 | 6 | 7 | 8 | 9 | 10 | 11 | 12 | 13 | 14 | 15 | 16 | 17 | 18 | 19 | 20 | 21 | 22 |
| Abrams Reactive Armor Tiles | 1 | RTBF | | | | | | 0.065 | 0.016 | | | | | | | | | | | | | | | |
| AFATDS | 2 | EMD | | | | | | | | | | | | 0.020 | | | | | | 0.050 | | | | |
| Air & Missile Defense Workstation | 3 | EMD | | | | | 0.012 | | 0.007 | | | | | 0.006 | 0.000 | 0.009 | 0.000 | | | | | | | |
| AN/GSR-8 (V)2 | 4 | RTBF | | | | | 0.015 | 0.000 | 0.006 | 0.000 | 0.025 | 0.000 | | 0.002 | 0.005 | | | 0.007 | | 0.014 | | | | |
| AN/PAS-13B Thermal sights | 5 | RTBF | | | | | | | | | | | | 0.002 | 0.000 | 0.000 | 0.000 | | | | | | | |
| Arcadia | 6 | RTBF | | | | | | | | | | | | 0.007 | 0.009 | 0.012 | 0.000 | 0.006 | | | | | | |
| Arcadia-Cerebus | 7 | RTBF | | | | | | | | | | | | | 0.000 | | | 0.005 | | 0.019 | | | | |
| Arcadia-TRSS | 8 | RTBF | | | | | | | | | | | | | 0.007 | | | 0.005 | | 0.009 | | | | |
| Area route clearing teams | 9 | RTBF | | | | | | | | | | | | | | | | | | | 0.021 | 0.012 | | |
| Army Non Lethal Capability Set | 10 | RTBF | | | | | | | | | | | 0.050 | | | | | | | | | | | |
| ASAS Light | 11 | RTBF | | | | | | | | | | | | | | | | 0.010 | 0.018 | 0.036 | | | | |
| ATACMS | 12 | EMD | | | | | | | | | | | | 0.022 | 0.000 | 0.000 | 0.100 | 0.020 | | | | | | |
| Automated Route Reconnaissance Kit | 13 | RTBF | | 0.012 | | | | | | | | | | | | | | | | | | | | |
| Backstop Systems | 14 | RTBF | | | | | | 0.025 | | | | | | | 0.016 | | | | | | | | | |
| Banshee | 15 | RTBF | | | | | | | | | | | | | | | | | | | | 0.024 | | |
| BFIST | 16 | RTBF | | | | | 0.036 | | | | | | | 0.000 | 0.000 | 0.000 | 0.067 | | | | | | | |
| Biometric Automated Tool Set | 17 | RTBF | | | | | | | | | | | | 0.000 | 0.040 | 0.000 | 0.000 | | | | | | | |

**Table A.2**
**Expected Values of Systems 18 to 34**

| System Name | Project Number | Developmental Stage | Gap Number | | | | | | | | | | | | | | | | | | | | | |
|---|---|---|---|---|---|---|---|---|---|---|---|---|---|---|---|---|---|---|---|---|---|---|---|---|
| | | | 1 | 2 | 3 | 4 | 5 | 6 | 7 | 8 | 9 | 10 | 11 | 12 | 13 | 14 | 15 | 16 | 17 | 18 | 19 | 20 | 21 | 22 |
| Blasting Device | 18 | RTBF | | | 0.100 | | | | | | | | | | | | | | | | 0.021 | 0.012 | | |
| Boomerang III | 19 | RTBF | | | | | | | | | | | | | | | | | | | | 0.013 | | |
| Bradley Reactive Armor Tiles | 20 | RTBF | | | | | | 0.065 | 0.016 | | | | | | | | | | | | | | | |
| Buffalo Mine Protected Clearance Vehicle | 21 | EMD | 0.000 | 0.019 | 0.025 | 0.028 | | | | | | | | | | | | | | | | | | |
| Bugler | 22 | RTBF | | | | | | | | | | | | | | | | | | | | 0.024 | | |
| Combat Periscope | 23 | RTBF | | | | | | | | | | | | | | | | | | | | | | |
| Combat Survivor Evader Locator | 24 | RTBF | | | | | | | | | | | | | | | | | | | | | 0.250 | |
| Convoy Auto Pilot (RAILCAR) | 25 | RTBF | | | | | | | | | | | | | | | | | | | | | | |
| Convoy Protection Platform | 26 | RTBF | | | | | | | 0.011 | | | | | | | | | | | | | 0.086 | | |
| Cougar/JERRV Medium Mine Protected Vehicle | 27 | RTBF | 0.000 | 0.019 | 0.025 | 0.028 | | | | | | | | | | | | | | | | | | |
| Counter Rockets, Artillery and Mortars (C-RAM) | 28 | RTBF | | | | | 0.099 | | 0.034 | | | | 0.000 | 0.033 | 0.000 | 0.034 | 0.000 | 0.025 | | | | | | |
| Crew Rescue brackets | 29 | RTBF | | | | | | | | 0.167 | | | | | | | | | | | | | | |
| CREW-2 | 30 | RTBF | | | | | | 0.000 | 0.042 | 0.000 | 0.000 | 0.000 | | | | | | | | | 0.042 | 0.024 | | |
| CROWS | 31 | RTBF | | | | | | | | | | | | | | | | | | | | | | |
| CROWS-Lite | 32 | RTBF | | | | | | | | | | | | | | | | | | | | | | |
| Dallas | 33 | RTBF | | | | | | | | | | | 0.056 | | | | | | | | | | | |
| DCGS-A | 34 | EMD | | | | | | | | | | | | | | | | 0.031 | | 0.067 | | | | |

**Table A.3**
**Expected Values of Systems 35 to 51**

| System Name | Project Number | Developmental Stage | Gap Number | | | | | | | | | | | | | | | | | | | | | |
|---|---|---|---|---|---|---|---|---|---|---|---|---|---|---|---|---|---|---|---|---|---|---|---|---|
| | | | 1 | 2 | 3 | 4 | 5 | 6 | 7 | 8 | 9 | 10 | 11 | 12 | 13 | 14 | 15 | 16 | 17 | 18 | 19 | 20 | 21 | 22 |
| Debris Blower (NASCAR) | 35 | RTBF | | | | | | | | | | | | | | | | | | | | | | |
| Defense Advanced GPS Receiver | 36 | RTBF | | | | | | | | | | | | | | | | | | | 0.000 | 0.007 | | |
| Delta Scientific DSC 1100 | 37 | RTBF | | | | | | | | | | | | 0.000 | 0.027 | 0.000 | 0.000 | | | | | | | |
| Dismounted 120mm Mortar Fire Control System | 38 | RTBF | | | | | | | | | | 0.333 | | | | | 0.067 | | | | | | | |
| Double Shot | 39 | RTBF | 0.000 | 0.006 | 0.000 | 0.000 | | | | | | | | | | | | | | | | | | |
| Double Vision | 40 | RTBF | | | | | | | | | 0.050 | | | | | | | | | | | | | |
| Driver Vision Enhancer | 41 | RTBF | | | | | | | | | | | | | | | | | | | | | | |
| Effects Management Tool | 42 | RTBF | | | | | | | 0.021 | | | | | | | | | | | | | | | |
| EM61-MK 2 | 43 | RTBF | | | | | | | | | | | | | | | | | | | | | | |
| Enhanced Logistical Support Off-Road Vehicle | 44 | RTBF | | | | | | 0.000 | 0.000 | 0.250 | 0.017 | 0.000 | | | | | | | | | | | | |
| Enhanced Mobile Raid | 45 | NEMD | | | | | | | | | | | | 0.007 | 0.002 | 0.007 | 0.000 | 0.009 | | | | | | |
| Enhanced-Tactical Automated Security System | 46 | RTBF | | | | | | | 0.007 | | | | | | | | | 0.005 | | | | | | |
| EOF Kit | 47 | RTBF | | | | | | 0.000 | 0.000 | 0.000 | 0.000 | 0.000 | 0.052 | | | | | | | | | | | |
| Escape Air – Emergency Breathing System | 48 | RTBF | | | | | | | | | | | | | | | | | | | | | | |
| Excalibur | 49 | EMD | | | | | | | | | | | | 0.022 | 0.000 | 0.000 | 0.067 | 0.020 | | | | | | |
| Expray | 50 | RTBF | | | | | | | 0.003 | | | | | 0.000 | 0.010 | 0.010 | 0.000 | 0.003 | | | | | | |
| Family of Medium Tactical Vehicles–Force Protection Kit | 51 | RTBF | | | | | | 0.020 | 0.005 | 0.000 | 0.000 | 0.000 | | | | | | | | | | | | |

## Table A.4
## Expected Values of Systems 52 to 68

| System Name | Project Number | Developmental Stage | Gap Number | | | | | | | | | | | | | | | | | | | | | |
|---|---|---|---|---|---|---|---|---|---|---|---|---|---|---|---|---|---|---|---|---|---|---|---|---|
| | | | 1 | 2 | 3 | 4 | 5 | 6 | 7 | 8 | 9 | 10 | 11 | 12 | 13 | 14 | 15 | 16 | 17 | 18 | 19 | 20 | 21 | 22 |
| FBCB2-BFT | 52 | EMD | | | | | | | | | | | | 0.037 | 0.000 | 0.094 | 0.000 | 0.010 | 0.015 | 0.030 | | | | |
| Fiberscope | 53 | RTBF | | | | | | | | | | | | 0.000 | 0.010 | 0.010 | 0.000 | 0.003 | | | | | | |
| FIDO Handheld Explosives Detection | 54 | RTBF | 0.010 | 0.010 | 0.050 | 0.000 | | | | | | | | | 0.020 | | | | 0.017 | | | | | |
| Flash Gordon | 55 | RTBF | | | | | | | | | | | 0.008 | | | | | | | | | | | |
| FN 303 Less Lethal Launcher | 56 | NEMD | | | | | | | | | | | 0.028 | | | | | | | | | | | |
| Forward Area Air Defense Command and Control | 57 | EMD | | | | | 0.024 | | | | | | | 0.006 | 0.000 | 0.009 | 0.000 | | | | | | | |
| FS3 Knight | 58 | RTBF | | | | | | | | | | | | | | | | | | | | | | |
| GMLRS | 59 | EMD | | | | | | | | | | | | 0.022 | 0.000 | 0.000 | 0.100 | 0.040 | | | | | | |
| Gotham | 60 | RTBF | 0.010 | | | | | | | | | | | | | | | | | | | | | |
| Green Laser Pointers | 61 | RTBF | | | | | | | | | | | 0.015 | | | | | | | | | | | |
| Ground Stand-off Mine Detection System Meerkats | 62 | EMD | 0.050 | 0.048 | 0.000 | 0.000 | | | | | | | | | | | | | | | 0.063 | 0.035 | | |
| Gunner Protection Kits | 63 | RTBF | | | | | | 0.043 | 0.011 | 0.000 | 0.000 | 0.000 | | | | | | | | | | | | |
| HazMATID systems | 64 | RTBF | | | | | | | | | | | | | | | | | 0.033 | | | | | |
| Hellfire 114K/P | 65 | RTBF | | | | | | | | | | | | 0.033 | 0.000 | 0.000 | 0.100 | | | | | | | |
| Highlighter | 66 | RTBF | | 0.018 | | | | | | | | | | | | | | | | | | | | |
| HMMWV Egress Assistance Trainer | 67 | RTBF | | | | | | 0.000 | 0.000 | 0.083 | 0.000 | 0.000 | | | | | | | | | | | | |
| HMMWV Frag Kit 1, 2, and 5 | 68 | RTBF | | | | | | 0.043 | 0.011 | 0.000 | 0.000 | 0.000 | | | | | | | | | 0.000 | 0.029 | | |

## Table A.5
## Expected Values of Systems 69 to 85

| System Name | Project Number | Developmental Stage | Gap Number | | | | | | | | | | | | | | | | | | | | | |
|---|---|---|---|---|---|---|---|---|---|---|---|---|---|---|---|---|---|---|---|---|---|---|---|---|
| | | | 1 | 2 | 3 | 4 | 5 | 6 | 7 | 8 | 9 | 10 | 11 | 12 | 13 | 14 | 15 | 16 | 17 | 18 | 19 | 20 | 21 | 22 |
| Hostile Artillery Locating System | 69 | RTBF | | | | | 0.068 | | | | | | | 0.025 | 0.000 | 0.021 | 0.000 | | | | | | | |
| Hunter | 70 | RTBF | 0.000 | 0.029 | 0.000 | 0.000 | 0.006 | | | | | | | 0.025 | 0.000 | 0.000 | 0.000 | 0.037 | | | | | | |
| Iceberg | 71 | RTBF | | | | | | | 0.003 | | | | | | | | | | | | | | | |
| IED Countermeasures Equipment (ICE) | 72 | RTBF | | | | | | 0.000 | 0.028 | 0.000 | 0.000 | 0.000 | | | | | | | | | 0.042 | 0.024 | | |
| I-GNAT | 73 | RTBF | 0.000 | 0.012 | 0.000 | 0.000 | 0.006 | | | | | | | 0.007 | 0.000 | 0.012 | 0.000 | 0.010 | | | | | | |
| Infrared Target Pointer AN/PEQ-2A | 74 | RTBF | | | | | | | | | | | | | | | | | | | | | | |
| Integrated Base Defense Security System | 75 | RTBF | | | | | | | | | | | | | | | | 0.025 | | | | | | |
| Interim Vehicle Mounted Mine Detection System (Husky) | 76 | RTBF | 0.050 | 0.048 | 0.200 | 0.074 | | | | | | | | | | | | | | | | | | |
| Intrepid Tiger | 77 | RTBF | | | | | | 0.000 | 0.028 | 0.000 | 0.100 | 0.000 | | | | | | | | | 0.000 | 0.000 | | |
| JLW155 | 78 | EMD | | | | | | | | | | | | 0.011 | | | | | | | 0.000 | 0.000 | | |
| Joint Air to Ground Missile | 79 | NEMD | | | | | | | | | | | | | | | | | | | | | | |
| Joint Blue Force Situational Awareness | 80 | RTBF | | | | | | | | | | | | | | | | | | | 0.000 | 0.020 | | |
| Joint Combat Identification Marking System | 81 | RTBF | | | | | | | | | | | | | | | | | | | | | | |
| Joint Warning and Reporting Network | 82 | EMD | | | | | | | | | | | | | | | | | 0.150 | | | | | |
| Kiddie Technologies Automatic Fire | 83 | RTBF | | | | | | | | | | | | | | | | | | | | | | |
| Land Warrior | 84 | EMD | | | | | | | | | | | | | | | | | | 0.050 | | | | |
| Land-based Phalanx (LPWS) G-8 | 85 | RTBF | | | | | | | 0.042 | | | | | | | | | | | | | | | |

**Table A.6**
**Expected Values of Systems 86 to 102**

| System Name | Project Number | Developmental Stage | Gap Number | | | | | | | | | | | | | | | | | | | | | |
|---|---|---|---|---|---|---|---|---|---|---|---|---|---|---|---|---|---|---|---|---|---|---|---|---|
| | | | 1 | 2 | 3 | 4 | 5 | 6 | 7 | 8 | 9 | 10 | 11 | 12 | 13 | 14 | 15 | 16 | 17 | 18 | 19 | 20 | 21 | 22 |
| Light Kit Motion Detector | 86 | NEMD | | | | | | | 0.003 | | | | | 0.000 | 0.013 | 0.010 | 0.000 | 0.003 | | 0.005 | | | | |
| Lightweight Counter Mortar Radar (Q-48) | 87 | EMD | | | | | 0.012 | | | | | | | 0.006 | | | | 0.005 | | | | | | |
| Lightweight Laser Designator/Rangefinder | 88 | EMD | | | | | | | | | | | | 0.005 | 0.000 | 0.000 | 0.000 | | | | | | | |
| Long-Term Armor Strategy | 89 | RTBF | | | | | | 0.024 | 0.006 | 0.000 | 0.000 | 0.000 | | | | | | | | | 0.000 | 0.029 | | |
| LVUSS | 90 | RTBF | | | | | | | | | | | | | 0.020 | | | | | | | | | |
| M113 Add on Armor (AoA) | 91 | RTBF | | | | | | 0.043 | 0.011 | | | | | | | | | | | | | | | |
| M141 BDM | 92 | RTBF | | | | | | | | | | | | | | | | | | | | | | |
| M2 | 93 | EMD | | | | | | | | | | | | | | | | | | | | | | |
| M249 (Squad Automatic Weapon) | 94 | RTBF | | | | | | | | | | | | | | | | | | | | | | |
| M4 Carbine | 95 | RTBF | | | | | | | | | | 0.167 | | | | | | | | | | | | |
| M4 Enhanced rail system | 96 | RTBF | | | | | | | | | | | | | | | | | | | | | | |
| M72A3/A7 | 97 | RTBF | | | | | | | | | | | | | | | | | | | | | | |
| M84 Diversionary Flash Grenade | 98 | RTBF | | | | | | | | | | | 0.008 | | | | | | | | | | | |
| M93 Fox Upgrade | 99 | EMD | | | | | | 0.000 | 0.009 | | | | | | | | | | | | | | | |
| MARCBOT | 100 | RTBF | 0.010 | 0.010 | 0.050 | 0.037 | | | | | | | | | | | | | | | | | | |
| Medium Directional Energetic Tool | 101 | RTBF | | | | | | | | | | | | | | | | | | | | | | |
| Metal Revetment Walls | 102 | RTBF | | | | | | 0.018 | | | | | | | 0.011 | | | | | | | | | |

## Table A.7
## Expected Values of Systems 103 to 119

| System Name | Project Number | Developmental Stage | Gap Number | | | | | | | | | | | | | | | | | | | | | |
|---|---|---|---|---|---|---|---|---|---|---|---|---|---|---|---|---|---|---|---|---|---|---|---|---|
| | | | 1 | 2 | 3 | 4 | 5 | 6 | 7 | 8 | 9 | 10 | 11 | 12 | 13 | 14 | 15 | 16 | 17 | 18 | 19 | 20 | 21 | 22 |
| MIDES | 103 | RTBF | | | | | | | | | | | | | | | | | | | | | | |
| Mine Resistant Ambush Protected Vehicle | 104 | RTBF | 0.000 | 0.019 | 0.025 | 0.028 | | | | | | | | | | | | | | | | | | |
| Mine Roller (Sharp Edge) | 105 | RTBF | | 0.008 | | | | | | | | | | | | | | | | | 0.000 | 0.013 | | |
| Mirage 1200 | 106 | RTBF | | | | | | | | | | | | | | | | 0.003 | | | | | | |
| Mk19 (40mm Machine Gun) | 107 | RTBF | | | | | | | | | | | | | | | | | | | | | | |
| Mobile Detection Assessment Response System MDARS | 108 | EMD | | | | | | | | | | | 0.000 | | | | | 0.003 | 0.017 | 0.005 | | | | |
| Mobile Vehicle Inspection System MVIS | 109 | RTBF | | | | | | | | | | | | 0.000 | 0.020 | 0.010 | 0.000 | | | | | | | |
| Mobi-Mat helicopter pads | 110 | RTBF | | | | | | | | | 0.033 | | | | | | | | | | | | | |
| Modular Launcher Communications System | 111 | RTBF | | | | | | | | | | | | | | | | | | 0.050 | | | | |
| Mortar Fire Control System | 112 | RTBF | | | | | | | | | | | | | | | | | | 0.050 | | | | |
| Movement Tracking System | 113 | EMD | | | | | | | | | | | 0.000 | | | | | | | 0.050 | | | | |
| Moving and Stationary Target Acquisition & Recognition | 114 | RTBF | | | | | | | 0.007 | | | | | | | | | | | | | | | |
| Neutralizing IEDs w/RF | 115 | RTBF | 0.000 | 0.008 | 0.000 | 0.000 | | | | | | | | | | | | | | | 0.021 | 0.024 | | |
| NL 40mm Multi-Grenade Launcher/Munition | 116 | RTBF | | | | | | | | | | | 0.008 | | | | | | | | | | | |
| NS Microwave Tactical Surveillance System | 117 | RTBF | | | | | | | | | | | | 0.005 | 0.011 | 0.012 | 0.000 | | | 0.009 | | | | |
| OAV | 118 | RTBF | 0.033 | 0.048 | 0.000 | 0.000 | 0.011 | | | | | | | 0.000 | 0.000 | 0.021 | 0.000 | 0.003 | | | | | | |
| ODIS | 119 | RTBF | 0.021 | 0.020 | 0.000 | 0.000 | | | | | | | | | | | | | | | | | | |

**Table A.8**
**Expected Values of Systems 120 to 136**

| System Name | Project Number | Developmental Stage | Gap Number | | | | | | | | | | | | | | | | | | | | | |
|---|---|---|---|---|---|---|---|---|---|---|---|---|---|---|---|---|---|---|---|---|---|---|---|---|
| | | | 1 | 2 | 3 | 4 | 5 | 6 | 7 | 8 | 9 | 10 | 11 | 12 | 13 | 14 | 15 | 16 | 17 | 18 | 19 | 20 | 21 | 22 |
| Omen 2.0 | 120 | RTBF | | | | | | | | | | | | 0.005 | 0.011 | 0.018 | 0.000 | 0.007 | | | | | | |
| Omnisense | 121 | RTBF | | | | | 0.015 | 0.000 | 0.006 | 0.000 | 0.025 | 0.000 | | 0.025 | 0.011 | | | 0.007 | | 0.014 | 0.000 | 0.020 | | 0.167 |
| Overwatch | 122 | RTBF | | | | | 0.055 | | | | | | | 0.000 | 0.000 | 0.000 | 0.000 | | | | | | | |
| PackBots | 123 | RTBF | 0.006 | | | | | | | | | | | | | | | | | | | | | |
| Persistent Surveillance and Dissemination System of System | 124 | RTBF | | | | | | | | | | | | 0.033 | 0.009 | 0.034 | 0.000 | 0.034 | | | | | | |
| Persistent Threat Detection System | 125 | RTBF | | | | | | | | | | | | 0.007 | 0.002 | 0.007 | 0.000 | 0.009 | | | | | | |
| Pitkin | 126 | RTBF | | 0.008 | | | | | | | | | | | | | | | | | | | | |
| Poltergeist | 127 | RTBF | 0.000 | 0.006 | 0.000 | 0.000 | | | | | | | | | | | | | | | | | | |
| Portable Barriers | 128 | RTBF | | | | | | | | | | | | | 0.027 | | | | | | | | | |
| PowerFlare Electronic Beacons | 129 | RTBF | | | | | | | | | | | 0.015 | | | | | | | | | | | |
| PPS-5D Ground Surveillance Radar | 130 | RTBF | | | | | | | | | | | | 0.002 | 0.005 | 0.012 | 0.000 | | | | | | | |
| Precision Lightweight GPS Receiver | 131 | RTBF | | | | | | | | | | | | | | | | | | | 0.000 | 0.007 | | |
| Profiler | 132 | RTBF | | | | | | | | | | | | 0.000 | 0.000 | 0.006 | 0.000 | | | | | | | |
| Prophet Spiral I ES | 133 | RTBF | | | | | | | | | | | 0.000 | | | | | | | | 0.000 | 0.013 | | |
| Prophet TROJAN LITE | 134 | RTBF | | | | | | | | | | | | 0.002 | 0.000 | 0.012 | 0.000 | | | | | | | |
| Publish and Subscribe System | 135 | RTBF | | | | | | | | | | | | | | | | | | 0.033 | | | | |
| Q36 | 136 | EMD | | | | | 0.012 | | | | | | | 0.006 | | | | 0.005 | | | | | | |

**Table A.9**
**Expected Values of Systems 137 to 153**

| System Name | Project Number | Developmental Stage | Gap Number | | | | | | | | | | | | | | | | | | | | | |
|---|---|---|---|---|---|---|---|---|---|---|---|---|---|---|---|---|---|---|---|---|---|---|---|---|
| | | | 1 | 2 | 3 | 4 | 5 | 6 | 7 | 8 | 9 | 10 | 11 | 12 | 13 | 14 | 15 | 16 | 17 | 18 | 19 | 20 | 21 | 22 |
| Ranger II | 137 | RTBF | | | | | | | 0.006 | | | | | | | | | | | | | | | |
| Rapid Access and Neutralization Tool | 138 | RTBF | 0.021 | 0.020 | 0.000 | 0.000 | | | | | | | | | | | | | | | | | | |
| Rapid Aerostat Initial Deployment | 139 | RTBF | | | | | | | | | | | | 0.007 | 0.002 | 0.007 | 0.000 | 0.009 | | | | | | |
| Rapid Deployment Integrated Surveillance System | 140 | RTBF | 0.000 | 0.012 | 0.000 | 0.000 | 0.015 | | | | | | | 0.005 | 0.011 | 0.018 | 0.000 | | | | | | | |
| Rapid Entry Vehicle | 141 | RTBF | | | | | | | | | | | | | | | | | | | | | | |
| Rapiscan Baggage Screeners | 142 | RTBF | | | | | | | | | | | | | 0.030 | | | | | | | | | |
| Raven Tactical MAV | 143 | RTBF | 0.000 | 0.012 | 0.000 | 0.000 | 0.011 | | | | | | | 0.007 | 0.000 | 0.012 | 0.000 | 0.010 | | | | | | |
| Razorback | 144 | RTBF | | 0.004 | | | | | | | | | | | | | | | | | | | | |
| Rearview (HMMWV Turret Mounted Mirrors) | 145 | RTBF | | | | | | | | | | | | | | | | | | | | | | |
| Recognition of Combat Vehicles (ROC-V) Training Software | 146 | RTBF | | | | | | | | | | | | | | | | | | | | | | |
| RG-31 Medium Mine Protective Vehicle | 147 | RTBF | 0.000 | 0.019 | 0.025 | 0.028 | | | | | | | | | | | | | | | | | | |
| Rhino Bus | 148 | RTBF | | | | | | 0.043 | 0.011 | 0.000 | 0.000 | 0.000 | | | | | | | | | | | | |
| Rhino II | 149 | RTBF | 0.000 | 0.010 | 0.000 | 0.000 | | 0.000 | 0.000 | 0.000 | 0.000 | 0.000 | | | | | | | | | | | | |
| Road Spikes – Magnum | 150 | RTBF | | | | | | | | | | | | | 0.013 | | | | | | | | | |
| Robber | 151 | RTBF | | 0.006 | | | | | | | | | | | | | | | | | | | | |
| Route Clearance Vehicle RCV EFP Up-Armor | 152 | RTBF | | | | | | 0.043 | 0.011 | 0.000 | 0.000 | 0.000 | | | | | | | | | 0.000 | 0.057 | | |
| Scorpion | 153 | RTBF | | | | | 0.023 | 0.000 | 0.003 | 0.000 | 0.000 | 0.000 | | 0.012 | 0.000 | | | 0.003 | | 0.009 | 0.000 | 0.007 | | 0.083 |

**Table A.10**
**Expected Values of Systems 154 to 170**

| System Name | Project Number | Developmental Stage | Gap Number | | | | | | | | | | | | | | | | | | | | | |
|---|---|---|---|---|---|---|---|---|---|---|---|---|---|---|---|---|---|---|---|---|---|---|---|---|
| | | | 1 | 2 | 3 | 4 | 5 | 6 | 7 | 8 | 9 | 10 | 11 | 12 | 13 | 14 | 15 | 16 | 17 | 18 | 19 | 20 | 21 | 22 |
| Secure 1000 | 154 | RTBF | | | | | | | | | | | | | 0.030 | | | | | | | | | |
| Sentinel CM | 155 | EMD | | | | | 0.012 | | | | | | | 0.006 | 0.000 | 0.009 | 0.000 | | | | | | | |
| Shades (IR Headlight Covers) | 156 | RTBF | | | | | | | | | | | | | | | | | | | | | | |
| Shadow | 157 | RTBF | 0.000 | 0.012 | 0.000 | 0.000 | 0.006 | | | | | | | 0.007 | 0.000 | 0.012 | 0.000 | 0.010 | | | | | | |
| Sniper Defeat | 158 | RTBF | 0.017 | | | | | | | | | | | | | | | | | | | | | |
| Special Weapons Observation Reconnaissance Detection Systems | 159 | RTBF | | | | | | | | | | | | | | | | | | | | | | |
| SPIDER | 160 | RTBF | 0.000 | 0.012 | 0.000 | 0.000 | | | | | | | | | 0.011 | | | | | | | | | |
| Standoff Daylight Warning Device | 161 | RTBF | | | | | | | | | | | | | 0.013 | | | 0.015 | | | | | | |
| Stryker FSV | 162 | RTBF | | | | | | | | | | | | | | | | | | | | | | |
| Stryker modifications | 163 | RTBF | | | | | | 0.000 | 0.021 | | | | | | | | | | | | | | | |
| Stryker Reactive Armor | 164 | RTBF | | | | | | 0.065 | 0.016 | | | | | | | | | | | | | | | |
| Talon – EOD Extended Reach Capability | 165 | RTBF | 0.010 | | | | | | | | | | | | | | | | | | | | | |
| Temporary Roadblock Apparatus Pack | 166 | RTBF | | | | | | | | | | | | 0.000 | 0.027 | 0.000 | 0.000 | | | | | | | |
| TOPSCENE | 167 | RTBF | 0.010 | 0.010 | 0.000 | 0.000 | | | | | | | | | | | | | | | | | | |
| Traffic Signs | 168 | RTBF | | | | | | | | | | | | | 0.013 | | | | | | | | | |
| Trailblazer | 169 | RTBF | | | | | | | | | | | | | | | | | | | | | | |
| Transparent gun shield | 170 | RTBF | | | | | | | | | | | | | | | | | | | | | | |

**Table A.11**
**Expected Values of Systems 171 to 183**

| System Name | Project Number | Developmental Stage | Gap Number | | | | | | | | | | | | | | | | | | | | | |
|---|---|---|---|---|---|---|---|---|---|---|---|---|---|---|---|---|---|---|---|---|---|---|---|---|
| | | | 1 | 2 | 3 | 4 | 5 | 6 | 7 | 8 | 9 | 10 | 11 | 12 | 13 | 14 | 15 | 16 | 17 | 18 | 19 | 20 | 21 | 22 |
| Tremors | 171 | RTBF | | | | 0.111 | | | | | | | | | | | | | | | | | | |
| Tunnel Detection | 172 | RTBF | | | | 0.167 | | | | | | | | | | | | | | | | | | |
| Unattended Transient Acoustic Measurement and Signatures Intelligence (MASINT) System | 173 | RTBF | | | | | 0.045 | | | | | | | 0.007 | 0.002 | 0.007 | 0.000 | | | | | | | |
| Vehicle Optics Sensor System | 174 | RTBF | | 0.008 | | | | | | | | | | | | | | | | | | | | |
| Viper Strike | 175 | RTBF | | | | | | | 0.042 | | | | | 0.016 | 0.000 | 0.000 | 0.000 | 0.023 | | | | | | |
| Walk Through Body Scan | 176 | RTBF | | | | | | | 0.009 | | | | | 0.000 | 0.030 | 0.021 | 0.000 | 0.007 | | | | | | |
| Warlock | 177 | RTBF | | | | | | 0.000 | 0.028 | 0.000 | 0.000 | 0.000 | | | | | | | | | 0.042 | 0.024 | | |
| Warrior | 178 | EMD | 0.000 | 0.029 | 0.000 | 0.000 | 0.006 | | | | | | | 0.040 | 0.000 | 0.000 | 0.000 | 0.060 | | | | | | |
| WebTAS | 179 | RTBF | | | | | | | | | | | | 0.012 | 0.005 | 0.031 | 0.000 | | | | | | | |
| Wide Area Surveillance Thermal Imager | 180 | RTBF | | | | | | | 0.006 | | | | | | | | | | | | | | | |
| Wireless Audio Visual Emergency System | 181 | RTBF | | | | | 0.012 | | 0.007 | | | | | | | | | | | | | | | |
| X26E TASER | 182 | RTBF | | | | | | | | | | | 0.008 | | | | | | | | | | | |
| Z-Backscatter Van | 183 | RTBF | | | | | | | 0.009 | | | | | 0.000 | 0.030 | 0.021 | 0.000 | 0.007 | | | | | | |

# Bibliography


Army Logistics Management College, *Analysis of Alternatives*, Fort Lee, Va.: Army Logistics Management College, PM-2009-DL, June 7, 2005.

Brown, Bradford, *Operation of the Defense Acquisition System, Statutory and Regulatory Changes*, Kettering, Ohio: Defense Acquisition University, December 8, 2008.

Chairman of the Joint Chiefs of Staff (CJCS), *Joint Capabilities Integration and Development System*, CJCSI 3170.01F, May 1, 2007a.

———, *Operation of the Joint Capabilities Integration and Development System*, CJCSM 3170.01C, May 1, 2007b.

Chow, Brian G., Richard Silberglitt, and Scott Hiromoto, *Toward Affordable Systems: Portfolio Analysis and Management for Army Science and Technology Programs*, Santa Monica, Calif.: RAND Corporation, MG-761-A, 2009. As of February 10, 2012:
http://www.rand.org/pubs/monographs/MG761.html

Chow, Brian G., Richard Silberglitt, Scott Hiromoto, Caroline Reilly, and Christina Panis, *Toward Affordable Systems II: Portfolio Management for Army Science and Technology Programs Under Uncertainties*, Santa Monica, Calif.: RAND Corporation, MG-979-A, 2011. As of February 10, 2012:
http://www.rand.org/pubs/monographs/MG979.html

CJCS—*see* Chairman of the Joint Chiefs of Staff.

Damstetter, Donald, "Life Cycle Management Improvement Initiatives," Office of the Deputy Assistant Secretary of the Army for Plans, Programs, and Resources (DASA [PPR]) and Office of the Assistant Secretary of the Army for Acquisition, Logistics and Technology (OASA [ALT]), November 16, 2004.

DDR&E—*see* Director of Defense Research and Engineering.

Director of Defense Research and Engineering (DDR&E), *Defense Technology Objectives for the Joint Warfighting Science and Technology Plan (JWSTP)*, Arlington, Va., February 2006.

DoD—*see* U.S. Department of Defense.

Greenberg, Marc, and James Gates, *Analysis of Alternatives*, Fort Belvoir, Va.: Defense Acquisition University, April 2006.

Landree, Eric, Richard Silberglitt, Brian G. Chow, Lance Sherry, and Michael S. Tseng, *A Delicate Balance: Portfolio Analysis and Management for Intelligence Information Dissemination Programs*, Santa Monica, Calif.: RAND Corporation, MG-939-NSA, 2009. As of February 10, 2012:
http://www.rand.org/pubs/monographs/MG939.html

Office of the Assistant Secretary of Defense (Public Affairs), *Secretary of Defense Provides Guidance for Improved Operational Efficiencies,* June 4, 2010a.

———, *Fact Sheet: Savings and Efficiencies Initiative*, June 4, 2010b.

Office of the Under Secretary of Defense, Comptroller, *DoD Financial Management Regulation,* Volume 2B: *Budget Formulation and Presentation,* "Research, Development and Evaluation Appropriations," 7000.14-R, July 6, 2000, Chapter 5 (December 2010), last modified February 23, 2011. As of February 10, 2012:
http://www.dod.mil/comptroller/fmr/02b/

P.L. 111-23, Weapon Systems Acquisition Reform Act of 2009, "Title II, Acquisition Policy," May 22, 2009.

Silberglitt, Richard, and Lance Sherry, *A Decision Framework for Prioritizing Industrial Materials Research and Development*, Santa Monica, Calif.: RAND Corporation, MR-1558-NREL, 2002. As of February 10, 2012:
http://www.rand.org/pubs/monograph_reports/MR1558.html

Silberglitt, Richard, Lance Sherry, Carolyn Wong, Michael S. Tseng, Emile Ettedgui, Aaron Watts, and Geoffrey Stothard, *Portfolio Analysis and Management for Naval Research and Development*, Santa Monica, Calif.: RAND Corporation, MG-271-NAVY, 2004. As of February 10, 2012:
http://www.rand.org/pubs/monographs/MG271.html

U.S. Army, *Military Operations: Force Operating Capabilities*, Fort Monroe, Va.: Training and Doctrine Command, pamphlet 525-66, July 1, 2005.

———, *2006 Army Modernization Plan: Building, Equipping and Supporting the Modular Force*, 2006a. As of October 22, 2010:
http://www.army.mil/features/MODPlan/2006/

———, *Descriptive Summaries for Program Elements of the Research, Development, Test and Evaluation*, Army FY 2007 in President's Budget Submission, February 2006b.

U.S. Army, Assistant Secretary of the Army for Acquisition, Logistics and Technology, *2005 United States Army Weapon Systems*, March 23, 2005.

U.S. Department of Defense (DoD), *Acquisition: Army Transition of Advanced Technology Programs to Military Applications*, Arlington, Va.: Office of the Inspector General, D-2002-107, June 14, 2002. As of October 27, 2010:
http://www.dodig.mil/audit/reports/fy02/02-107.pdf

———, Deputy Secretary of Defense Gordon England, *Capability Portfolio Management Test Case Roles, Responsibilities, Authorities, and Approaches*, September 14, 2006.

———, Deputy Secretary of Defense Gordon England, *Capability Portfolio Management*, DoD Directive #7045.20, September 25, 2008.

———, Under Secretary of Defense for Acquisition, Technology and Logistics (USD [AT&L]), *Operation of the Defense Acquisition System*, DoD Instruction 5000.2, May 12, 2003. As of October 21, 2010:
http://www.carlisle.army.mil/dime/documents/JPLD_AY08_Lsn%207_Reading%203_DoDI%205000-2.pdf